Food Production in Native North America

An Archaeological Perspective

Kristen J. Gremillion

SOCIETY FOR AMERICAN ARCHAEOLOGY
The SAA Press
Washington, DC

The Society for American Archaeology, Washington, DC 20005
Copyright © 2018 by the Society for American Archaeology
All rights reserved. Published 2018

Library of Congress Cataloging-in-Publication Data

Names: Gremillion, Kristen J., 1958- author.
Title: Food production in native North America: an archaeological perspective / Kristen J. Gremillion.
Description: Washington, DC : The Society for American Archaeology, 2018. | Includes bibliographical references and index.
Identifiers: LCCN 2017058700 | ISBN 9780932839572 (print)
Subjects: LCSH: Indians of North America--Food. | Indians of North America--Agriculture. | North America--Antiquities.
Classification: LCC E98.F7 G73 2018 | DDC 970.004/97--dc23
LC record available at https://lccn.loc.gov/2017058700

Printed on acid-free paper

Contents

Acknowledgments		v
Food Production in Native North America: An Introduction		vii
1	A Coevolutionary Continuum	1
2	The Eastern Agricultural Complex	12
3	Origins and Development of Maize-Based Agriculture in the Southwest	38
4	The Rise of the Three Sisters: Maize in the Eastern Woodlands	65
5	Food Production without Farming	90
6	A World of Difference: Food Production in Postcontact North America	105
7	Synthesis	122
Appendix A: Scientific Names of Plant and Animal Taxa		145
References		150
Index		185

Acknowledgments

First I want to thank the Native peoples of North America, the first ecologists and stewards of the land. It is their knowledge and care across thousands of years that coaxed crops from the desert and taught the Three Sisters how to dance together. I also thank the archaeologists and archaeobotanists who brought to light and sought to understand the material traces of ancient food production systems. Ken Ames and Michelle Hegmon of SAA Press provided valuable guidance and encouragement along the way. Two anonymous reviewers offered helpful commentary that greatly improved the final product, and Paul Minnis graciously reviewed the chapter on the Southwest. Marnie Colton of The SAA Press efficiently guided me through the production process, from copyediting to publication.

A special thanks goes to my spouse Paul Gardner, for always being in my corner.

Food Production in Native North America: An Introduction

There was a time when North American archaeologists were content to differentiate between hunter-gatherers (who could only harvest what nature provided) and farmers (who had acquired agriculture from their more advanced Mesoamerican neighbors). Not so today, when food production is widely recognized as a broad strategy of intensification that takes a variety of forms, many of which diverge from the traditional model of Old World–style grain agriculture. In fact, the prehistoric past reveals a much more diverse range of subsistence adaptations. Even in regions long considered nonagricultural, such as the Northwest Coast and the Great Basin, subsistence intensification is now known to have considerable time depth. The practices involved, such as burning, soil modification, coppicing, and transplanting, do not conform to concepts of food production based on domesticated plants and animals.

In the North American midcontinent, initial food production took the form of small-scale cultivation of several seed-producing annual plants that began no later than 3000 BC. By about AD 1, this "Eastern Agricultural Complex" (EAC) had become a key component of a diversified foraging-farming economy. It was only after about AD 800–1000 that maize agriculture rose to prominence in eastern North America, at least a millennium after its initial introduction to the region. Maize, a crop of Mesoamerican origin, arrived in the Southwest prior to 2000 BC, around the time that husbandry of native crops was emerging in the East. Maize quickly acquired a role as a staple crop in many parts of the Southwest, despite the challenges of maintaining an adequate supply of moisture in this relatively dry climate. Southwestern farmers coped with aridity by developing sophisticated methods of water conservation and control and practicing careful site selection to take advantage of groundwater. However, wherever it developed, maize-centered farming remained diversified enough to buffer unpredictable crop losses with a roster of foraged foods and other domesticated plants, both native and introduced.

Despite this variability, some features distinguish North America as a whole from other regions in which food production took hold. Perhaps the most important of these is the absence of pastoralism prior to European contact. Without animal husbandry, there was no tradition of dairy production, and meat was obtained entirely through hunting and fishing (or trade). Nonagricultural plant foods such as tree nuts remained important sources of nutrition up until, and beyond, initial European contact. Why this pattern exists is not entirely clear, although it may represent a historical constraint, namely the scarcity of animal species with appropriate behavioral and ecological traits for domestication (Zeder 2015). Turkeys were domesticated in the Southwest and in some cases made substantial contributions to human diets (Munro 2011), but did not demand the same amount of labor or level of care associated with cattle, sheep, pigs, and goats in mixed farming systems of the temperate Old World. With domestic animals playing a minor role, the story of food production in North America is dominated by the relationships between human populations and plants (Smith 2011b:6).

One implication of the absence of animal husbandry is that nutritional needs (especially for protein) continued to be met by foraging activities well after domesticated plants had become staple foods. Hunting of large game such as deer likely also had social significance as an indicator of male accomplishment (Zeanah 2004). Lacking the predictability and degree of control afforded by animal husbandry, subsistence systems were particularly sensitive to fluctuations in wild animal populations. In the Eastern Woodlands, this vulnerability may explain the emphasis by farming communities on sustainable food sources such as deer and freshwater fish (Smith 2009).

In North America, the material technology of food production remained relatively simple throughout prehistory (a fact that does not imply simplicity of the associated knowledge systems and practices). Plows and draft animals were European introductions. Clearing, planting, and cultivating relied on edged stone tools, digging sticks of wood or bone, and fire. In the arid West, methods of water control were used that included mulching and passive irrigation, and maize farmers at relatively high latitudes created ridged fields to buffer the effects of frost during the growing season (Doolittle 2000). Peoples of the Pacific Northwest managed gardens of starchy tubers and roots and burned berry patches to increase production (Deur and Turner 2005).

Despite having been supported by a relatively simple material technology, prehistoric North American food production was informed by a complex and sophisticated body of cultural knowledge accumulated across millennia of residence on the continent. Such knowledge systems are, in some fortunate cases, still in existence and are likely to persist into the future. Across much of the continent, however, traditions have been irrevocably disrupted and weakened (though emphatically not eradicated) by European colonization. Especially because of the damage done to orally transmitted bodies of knowledge, the archaeological record—the material remnant of past human behavior—is in many cases all that remains of ancient subsistence practices. That record is essential to understanding why systems of food production arose, persisted, and changed in North America.

This book provides a highly selective survey of Native North American food production systems from an archaeological perspective. It has four major foci: the domestication and intensification of indigenous seed crops in the East; the introduction and spread of maize-based farming systems that incorporated crops of Mesoamerican origin, including maize; the persistence of diverse low-intensity forms of food production in societies that evade the classic forager-farmer dichotomy; and the impact of introduced crops after AD 1492. These topics are flanked by an introduction to the ecological and cultural variability of North America across space and time, and a concluding discussion of causal explanations that have been proposed for the development of food-producing socioeconomic systems in the region.

The theoretical framework used in this volume is both evolutionary and ecological. Natural selection is the driving force behind domestication; when humans modify the habitats of the plants and animals on which they depend, they nudge selection in new directions (deWet 1975; Rindos 1984; Smith 1987b). As mutual dependency increases, distinctive plant phenotypes develop that reflect human management practices and preferences. These preferences are products of our evolutionary history as a species, even though the decisions themselves conform to specific local conditions and reflect particular cultural values. This degree of flexibility does not imply that procurement, processing, and consumption of food occur randomly or according to whim; all reflect species-wide patterns of cognition and social interaction in environmental context. Attention to economic utility and effective social interactions (whether cooperative or competitive) has served our species well for millennia. Unlike other species, however, humans preserve advantageous strategies over multiple generations by means of social learning. This system permits ongoing adjustments as people innovate and incorporate the products of social and experiential learning (Henrich and McElreath 2003). Thus there is nothing contradictory about considering the evolutionary basis of human behavior, the influences of social learning, and individual agency as partners in the explanation of subsistence transitions. I make the assumption that economic utility always has some relevance for understanding these transitions because efficiency and risk have powerful effects on survival and the ability to influence others. This generalization remains true despite the fact that individuals do not always achieve, or even strive for, optimal solutions to economic trade-offs.

In North America, as elsewhere on the planet, human communities faced the challenge of feeding themselves while simultaneously meeting an array of other social and biological needs. Some populations came to rely on food production to help them meet this challenge; others did not. This book asks why this pattern exists, with the assistance of some foundational assumptions from evolutionary ecology and archaeological data from a wide range of environmental settings. It does not propose a single dominant causal factor or argue for identical historical pathways to food production; instead, it looks at those pathways as historical enactments of cultural, evolutionary, and ecological processes that affect human societies worldwide.

1
A Coevolutionary Continuum

During the twenty-first century, the study of the origins and development of food production has been revolutionized by methodological advances and a growing sophistication among archaeologists regarding ecological processes and theories. In this climate, the long-standing dichotomy between foraging and farming has eroded to the point of near-collapse. It has become widely recognized within the research community that the intensity of human interactions with plant and animal resources varies along multiple dimensions, rather than sorting cleanly into two, or even just a few, categories (Harris 1989, 2007; Smith 2001, 2005b, 2011c). When these interactions develop into mutualism, in which both populations receive fitness-enhancing benefits from the relationship, coevolution occurs (Rindos 1984). At one end of the coevolutionary continuum, the human behaviors that drive coevolution are either absent or limited in extent and impact and may be unintentional; at the other, humans play an active role in modifying plant habitats and life cycles, creating anthropogenic landscapes, and developing production strategies.

Along the continuum of human-plant interaction there are many points at which societies may reach a comfortable equilibrium and persist there for centuries or even millennia without crossing a clear threshold that would mark them unambiguously as farmers. These systems occupy the middle ground labeled "low-level food production" by Smith (2001, 2005b). For example, Northwest Coast groups, many of which practiced intensive management of plants that produce edible underground storage organs, occupy this part of the continuum. These peoples are in an important sense food producers, yet (unlike the farmers with whom they are inevitably contrasted) they manage economically important resources in situ instead of concentrating them in plots, corrals, and fields. Rather than moving the plants from their natural habitat, they modify the habitat itself, for example, by building soil-conserving walls or burning vegetation to encourage fruit production (Deur and Turner 2005). Because these food production systems focus on perennial plants (which persist across multiple growing seasons), they may be slow to produce the morphological markers used to identify domesticated forms of annual plants (which die and set seed annually).

In contrast, the annual harvesting and replanting required to raise annual seed-producing crops ensures that an opportunity exists to modify the gene pool at least once each year. The annual habit may explain why the native crops that came to occupy a position of economic importance in eastern North America so often show morphological markers of domestication. Annual seed crops were one element of a stable adaptation that combined foraging, collecting, and domestication of a select group of annuals—in Smith's (2005b) taxonomy, low-level food production with domestication. More intensive forms of agriculture with domesticates developed only late in prehistory and placed maize in a central role. In the face of such diversity, it seems simplistic to speak of a transition from foraging to farming. We need a better way to divide the continuum, however arbitrarily, for purposes of discussion.

Smith (2005b) has discussed these terminological issues at length and suggested a solution, which I adopt here in modified form. This usage both acknowledges the "middle ground" as a form of food production and recognizes that agriculture is a significant phenomenon that differs in important ways from other forms of resource management. Agriculture anchors one end of the continuum. The opposite end, "food procurement," is largely congruent with "foraging" or "hunting and gathering" in conventional terminology. I use "food production" broadly to include both Smith's middle ground of "low-level food production" and agriculture.

I use "agriculture" and "farming" to label subsistence systems and activities that 1) involve intensification (an increase in yields per unit land beyond that available from unmodified habitats); 2) rely to some degree on domesticates (animals or plants whose relationship with humans has altered selection pressures, creating population-level differences from wild and weedy forms; 3) contribute a major proportion of food energy for a human community or household, at least for part of the year; 4) require an investment of household labor that must be diverted from other activities (and thus entails some opportunity costs); and 5) create persistent anthropogenic habitats. An additional criterion to take into account is whether the activities in question are necessary to sustain the full range of social, ceremonial, and economic activities that are considered normative by community members. If present, evidence of cultural significance strengthens the case for agriculture; however, its absence should not be weighted heavily because of the difficulty of detecting ideology in the archaeological record.

Like human-plant interaction in general, specific plant taxa are best described as occupying places along a continuum rather than fitting neatly into discrete categories (Harris 1989; Smith 2001). At one end of the continuum lie the domesticates, which have adapted to human management by producing unusually high frequencies of phenotypic traits that would be deleterious in other ecological settings. Often these domestication-associated traits are reflected in the morphology of seeds and fruits and are easily recognized by a competent observer. More direct evidence comes from ancient genomes that record change at specific loci. However, even if domestication cannot be demonstrated in this way, in some cases there is other evidence suggesting human management in the form of planting, transplantation, weeding, protection, habitat enrichment, and other supportive activities. This evidence may come from biogeography (testifying to human-mediated range expansion), contextual association with known domesticates, or archaeological indicators of intensive harvesting, processing, and consumption. These I refer to as "cultivated plants" or "managed plants." Weeds as defined here are not plants that are unwanted; in ecological terms, a weed is a plant that is adapted to disturbed habitats, including those created by human activity. Weeds tend to be opportunistic colonizers of open habitats and thus share a suite of features including rapid growth and reproduction, high yields of seeds and fruits, and broad environmental tolerance (Baker 1974; deWet and Harlan 1975; Harlan and deWet 1965). Finally, I reserve the term "wild" to indicate plants at the end of the ecological continuum farthest from domesticates. Unlike weeds, wild plants show no particular affinity for human-modified habitats and may be poorly adapted to survive in them. Both wild plants and weeds are considered free-living, in contrast to the tethered existence of managed plants and domesticates.

The Continuum in Ecological Context

With these parameters in mind, certain regions of North America stand out as appropriate topics for a book on food production. I begin by identifying ecological settings in which agriculture was never a feasible way of making a living, generally because their environmental parameters exceed the tolerance limits of crop plants. In marginal zones, crop cultivation was possible but too unproductive to be viable, preventing agriculture from making significant incursions until (and unless) improvements occurred in technology or growing conditions. With these constraints out of the way, I focus on environments in which agriculture was possible but had varying success as a persistent subsistence economy.

Environmental Constraints

It goes without saying that the range of a plant is constrained by the environmental conditions that affect its growth and ability to reproduce. Although crop plants benefit from human interventions that can help them overcome some of these constraints, they cannot grow wherever people want them to. Plants indigenous to a region are adapted to the local climate, but not necessarily to the full range of habitats associated with it. Of particular relevance to the origins of food production is tolerance for ecological disturbance, which can open the door to a more fully realized coevolutionary relationship with humans. For example, several of the native seed crops of eastern North America were derived from weeds of floodplains and other naturally disturbed areas and thus were well adapted to habitats disturbed by human activity, such as campsites. They benefit from enriched soils but do not require them, and typically show broad environmental tolerance (Baker 1974). In contrast to these weedy crops, maize was a non-native plant with origins in the Balsas River valley of central Mexico (Matsuoka et al. 2002). Although its spread across the North American continent was slow compared to that of native crops, maize was grown successfully in extremely arid environments (sometimes with the aid of irrigation) and was able to adapt to the low winter temperatures and long summer days of high latitudes.

In the north, the chief constraint on plant development is the length of the growing season. Maize requires a minimum of 120 days to reach maturity (Hart 2014; Yarnell 1964), and if the growing season is near that minimum, early or late frosts can ruin a crop. Despite these risks, maize cultivation eventually reached the northern Great Plains and parts of southern Ontario, as did some of the native crops (Adair 2003; Boyd and Surette 2010; Crawford et al. 1998). At higher latitudes, shorter growing seasons ruled out agriculture. West of the Plains, growing maize often required strategies designed to direct, capture, and conserve moisture. Plants other than maize were cultivated that were tolerant of dry conditions, such as agave, and many more drought-tolerant wild plants served as fallback foods when maize and other cultigens produced poorly (Anderies et al. 2008; Minnis 1991). Despite human adaptability, there remained hyperarid localities in the West where crop cultivation was not an option. An additional challenge in the Southwest was the trade-off between moisture and temperature—higher elevations could often support rainfed agriculture, but at the risk of damaging frost (Cordell and McBrinn 2012). Throughout the rest of North America, environmental conditions permitted food production; whether people chose to practice it, and in what form, depended on numerous historical, ecological, and socioeconomic variables.

Major Ecological Regions

The portion of North America suitable for food production is vast and varied in its vegetation, topography, and climate. Patterns of vegetation become quite complex west of the Great Plains, where high mountains create extreme contrasts in vegetation and associated fauna. Deserts, Mediterranean scrub, woodlands, and temperate forests are all represented here (Figure 1.1). In the following section, I discuss this variation using a simplified version of the ecoregions of North America (Commission for Environmental Cooperation 1997, 2011), omitting those found primarily in Mesoamerica as well as northern ecoregions where food production was not generally practiced prehistorically. Human groups adapted successfully to all of these settings by practicing a wide range of cultural strategies for acquiring and managing their natural resources. Broadly similar adaptations to shared environments have been used to define culture areas (Smith 2011b), which are shown in Figure 1.2 along with some important physiographic features. Unless otherwise noted, ecoregion characteristics are drawn from the North American Environmental Atlas (Commission for Environmental Cooperation 2011) and Commission for Environmental Cooperation (1997).

EASTERN TEMPERATE FORESTS. The continent east of the Great Plains, south of the Great Lakes, and north of the southern tip of Florida supports

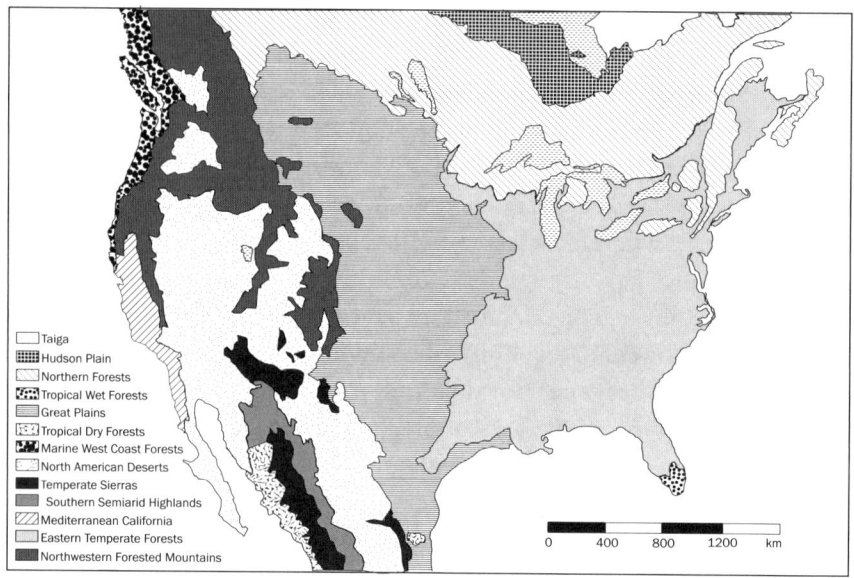

Figure 1.1. Map of North America showing major ecological regions. Adapted from Commission for Environmental Cooperation (1997, 2011).

temperate forests, most of which are dominated by deciduous trees (which shed their leaves annually). Depending on edaphic factors and disturbance regimes, conifers such as pines are also dominant in some areas, such as the southeastern Coastal Plain (Figure 1.2). At higher latitudes, northern hardwoods such as beech, maple, and basswood give way increasingly to evergreen conifers, eventually being replaced by northern forests of spruce, pine, fir, and tamarack in the Northern Forest ecoregion (Figure 1.1; see Appendix A for Linnaean taxa corresponding to vernacular names). The highly diverse mixed mesophytic forests of the Appalachian region are unusual in lacking consistently dominant species. Although valued today for their biodiversity (and much reduced in area), these mesic forests were less productive from the human perspective than the drier forest types in which oak is one of the dominant genera (Cowan 1985a). Oaks are codominant with pines across much of the Atlantic Coastal Plain (Figure 1.2) and make up a major component of mixed forests in the Appalachian summit region and along the northeastern coast (Dyer 2006). In the Piedmont foothills just east of the Appalachians, pines are a common constituent of mixed forests and dominate the sandy soils of the Coastal Plain. At the southern tip of Florida,

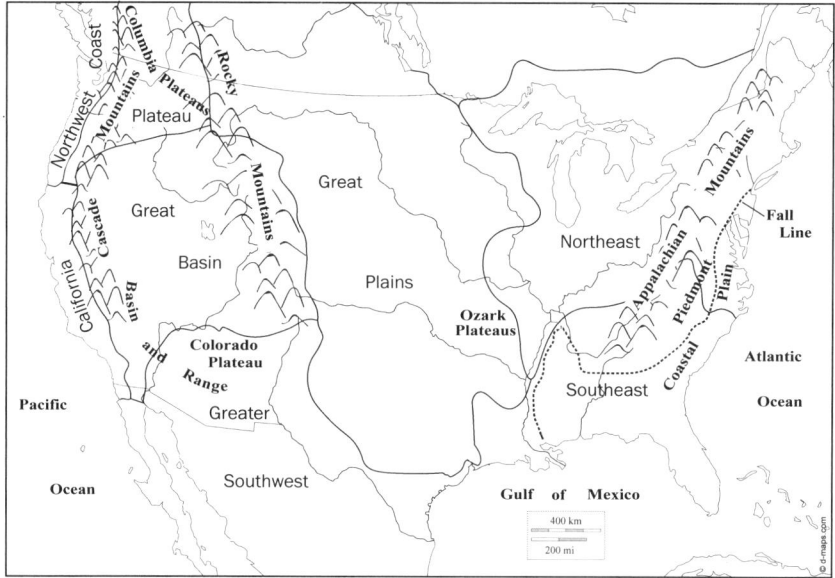

Figure 1.2. Map of North America showing broadly defined culture areas and important physiographic features. Adapted from Smith (2011b:4).

tropical conditions produce a wet evergreen forest with wetlands and mangrove swamps at lower elevations. Forest grades into parkland and grassland at the edge of the Great Plains.

Several forest trees were important sources of carbohydrate calories for people living in the region, including species of oak and hickory as well as butternut (white walnut), black walnut, and hazelnut (Gardner 1997). Other resources include whitetail deer (probably the most important game animal) along with smaller mammals such as raccoon, opossum, beaver, and eastern cottontail (Lapham 2011; Styles 2011). Other prey included turtles, migratory waterfowl, fish, and shellfish. Freshwater, saline, and brackish aquatic habitats in close proximity offered high biotic productivity and species richness along the coastal zone.

Farming in the Eastern Temperate Forests was supported by ample rainfall suitable for a wide range of crops. The region's many streams and valleys offered fertile, easily worked soils, and the adjacent uplands supported a variety of game animals. However, shaded forest floors had to be exposed to sunlight to create suitable habitats for rapid growth and high seed production in annual crops. Natural disturbances (e.g., wildfires, flooding, slope erosion) create openings in the forest canopy, but Native farmers found ways to mimic these conditions in desired locations. Historic groups often used controlled burning for this purpose, and the records provided by sediment cores indicate that this practice was an ancient one (Delcourt 2008).

GREAT PLAINS. The Great Plains occupy the continent's midsection. West of the Mississippi Valley, deciduous forests gradually give way to grassland with decreasing amounts of rainfall. Topography is somewhat varied, including a mix of smooth and rolling plains and some hills. Before Euro-American settlement, the dominant vegetation was grassland, ranging from tallgrass prairie and woodlands on the eastern edge to the shortgrass prairies of the drier western zone. Wetlands of glacial origin are particularly abundant in the northern plains and are important habitats for migratory waterfowl. Larger animal prey included bison, pronghorn antelope, elk, and mule deer.

The development of food production in the Great Plains followed a variety of pathways depending on environmental variables and communication with adjacent regions. In the eastern Plains, there is ample evidence of reliance on native seed-producing annual plants after 1500 BC. Farther west, this native crop complex did not gain a foothold. However, maize-based agricultural systems emerged across the region starting around 100 BC. Variants of this Plains Village tradition combined maize farming with cultivation of native crops in the eastern Plains and with buffalo hunting in the west (Adair 2003; Adair and Drass 2011).

WESTERN FORESTS. The Northwestern Forested Mountains extend from Alaska south into northern California and eastward to the Rocky Mountains. In this region, extremes of moisture and temperature can occur over short distances. Its vegetation is diverse, with alpine tundra on the highest peaks and semiarid grasslands and brush at lower elevations. Intermediate elevations support subalpine coniferous forests. Economically important mammals of the region include mule deer, elk, moose, mountain goat, bighorn sheep, and black and grizzly bear, which were often pursued in higher-elevation woodlands and parklands. Some groups who occupied these montane forests and adjacent deserts planted crops, but the economic importance of these resources varied widely (Fowler and Rhode 2011).

The Marine West Coast Forests ecoregion occupies the Pacific Coast from central California to Alaska. This region's mountains and valleys support temperate rainforests as well as boreal forests at higher elevations. Terrestrial fauna is similar to that found in other western forests, and the coast adds marine mammals, seabirds, and fish, including salmon. Maritime influence buffers temperature extremes and ensures a long growing season as well as providing ample rainfall. The forest soils are nutrient-poor, but major river valleys and some coastal areas are highly productive farmland.

The West Coast forests and adjacent inland zones supported substantial human populations that relied on seasonally abundant salmon and other anadromous fish along with a wide array of naturally occurring plant resources. The abundance and reliability of these foods is a frequently cited cause of the failure of northwestern peoples to develop or adopt agriculture. However, recent investigations of traditional botanical knowledge and practice have revealed substantial evidence of management of resources that mimics the boost in yields and predictability enjoyed by agriculturalists. These traditions represent a kind of food production long unrecognized by scholars accustomed to equating agriculture with annual seed crops such as cereals (Deur 2005; Smith 2005b).

NORTH AMERICAN DESERTS. Deserts are an important component of the arid West, occupying a large portion of the area between the Great Plains and the Pacific Coast and forming a mosaic with forests in mountainous areas. The climate of the western deserts varies from arid to semiarid, with strong seasonal contrasts in temperature and often rainfall as well. Winter moisture prevails in the northern parts of the region, with summer rains more characteristic of southern deserts. The deserts support a number of distinctive vegetation associations adapted to low rainfall and other environmental extremes. Sagebrush, creosote bush, Joshua tree, paloverde, and numerous species of cacti and agave are frequent dominants. High relief creates altitudinal zonation of vegetation along a moisture gradient that separates desert and steppe vegetation from upland forests. Precipitation ranges from 130 mm to 380 mm, the low end of which introduces significant challenges for food production, particularly for crops that are not desert adapted.

Despite these challenges, intensive agriculture was widespread in the deserts of North America. In the cold desert of the Great Basin, hunting-gathering lifeways persisted longer, although maize agriculture was adopted as a supplement by some groups living on its southern margin (Fowler and Rhode 2011). Native communities of the Greater Southwest have practiced traditional drylands agriculture for millennia under widely varying environmental conditions. These communities were the first in North America to acquire maize from Mesoamerica and initiated its introduction to the rest of the continent. Southwestern agricultural systems were attuned to sparse and variable rainfall regimes and made strategic use of multiple resource zones for agriculture, hunting, and collection of wild plant foods. At its most productive, the North American Deserts region supported substantial human populations. However, the aridity of the region made agriculturally based growth especially risky (Benson and Berry 2009).

MEDITERRANEAN CALIFORNIA. Much of the Pacific Coast has a Mediterranean climate, characterized by cool, wet winters and warm, dry summers. Coastal ranges and valleys create a topographically complex landscape surrounding a large, level central valley with deep soils. The frost-free period ranges from 250 to 350 days, an ample growing season for many crops that cannot be cultivated elsewhere in North America. Vegetation is mostly chaparral (evergreen shrub vegetation), with savannah-like oak woodlands, but many other plant communities coexist, often within short distances of each other (Bettinger and Wohlgemuth 2011). Prehistorically, mammals were the most important terrestrial animal foods, including elk or wapiti, mule deer, pronghorn antelope, bighorn sheep, and a variety of small game and terrestrial birds. On the coast, marine mammals were pursued where technology permitted (Hildebrandt and Carpenter 2011).

Native groups of California, like those of the Pacific Northwest, have long been of interest to economic anthropologists because they practiced resource intensification without domesticated plants or other ingredients of agriculture as it is traditionally conceived. The "acorn economies" of central California, for example, were based on harvesting of a wide variety of animal and plant foods, many of which were managed with controlled burning, coppicing, pruning, and other techniques (Anderson 2005; Bettinger and Wohlgemuth 2011). These practices were directed primarily at perennial plants, including the oak trees (*Lithocarpus* and *Quercus*) that were a key source of carbohydrate calories (Basgall 1987).

Summary

Humans entered the North American continent as large mammals with a powerful and flexible adaptive system in which accumulated knowledge and innovation both played important roles. Initially adapted to high-latitude habitats, early Americans dispersed southward to encounter a wide range of temperate and tropical ecosystems. In many regions, key locations attracted human residents who reduced their foraging ranges in favor of more persistent settlements. Residential stability encouraged the development of sophisticated and habitat-specific subsistence technologies. Those technologies expanded to include active interventions that in many cases improved foraging efficiency by increasing yields of plant foods and attracting game animals. Although such practices were geographically widespread, in most regions they stopped short of the cycle of planting and harvesting that results in domestication. Only in the eastern forests and adjacent plains were native plants recruited to form the basis of a small-scale but fully developed system of food production: the Eastern Agricultural Complex (EAC).

2

The Eastern Agricultural Complex

The importance of maize in the subsistence economies of Eastern Woodlands peoples was obvious to the first European chroniclers. At initial contact, maize-based agriculture supported the complex polities of the Southeast as well as the villages of the Northeast. This understanding of Native food production as largely dependent upon crops of Mesoamerican origin—including the common bean and squash[1] as well as maize—prevailed through much of the twentieth century. That this characterization was incomplete became apparent with the discovery and focused investigation of pre-maize cultivation of native plants. Weedy annuals such as pitseed goosefoot, marshelder, and an indigenous squash that had been brought under domestication thousands of years prior to the introduction of maize remained on the landscape only as free-living populations. Of these pre-maize crops, only the sunflower persisted in its domesticated form.

Since the mid-twentieth century, the understanding of these early systems of food production has grown considerably with the movement into the mainstream of such techniques as fine-screen recovery of plant remains (flotation), high-magnification microscopy, direct radiocarbon dating of botanical material, multivariate statistics, and ancient DNA (Popper 1988; Smith 1992b, 2006a; VanDerwarker 2010; Zeder et al. 2006). As a result of this body of work, eastern North America is today one of the most thoroughly documented cases of prehistoric agricultural origins and is acknowledged as an independent center of plant domestication (Smith 2006b). The system of food production based on native crops is sometimes termed the "Eastern Agricultural Complex" (EAC). Although this term has been contested because the members of the "complex" vary across time and space (Scarry and Yarnell 2011; Yarnell 1986), it has been revived with the discovery of very early evidence of the four major crops in association at the Riverton site in Indiana (Smith and Yarnell 2009). Usage of the term EAC does not imply that all its member species were cultivated together or transmitted as a package, but it does highlight the existence of a prior and historically distinctive system of food production that was in many ways quite different from the maize-based systems that largely replaced it.

Chronology and Culture History

In the discussion to follow, I use large-scale culture historical units to designate major chronological divisions (Figure 2.1). Although these periods have been traditionally associated with lists of cultural traits supposed to be diagnostic (e.g., appearance of first pottery), these temporal associations have begun to fray under the weight of new evidence. Pottery, for example, is no longer a marker for the Woodland period because its origins are now known to extend back into the Archaic period in some parts of the East. I ignore trait associations in this volume and use the terms in Figure 2.1 in a purely chronological sense. I generally avoid archaeological phase names in favor of corresponding region-wide units, but make exceptions when a well-understood local sequence provides convenient labels.

Plant domestication was well underway at several locations within eastern North America by approximately 3,000 BC (Smith 2011a), within the Late Archaic period as it is usually defined. Only after about 1000 BC do substantial deposits of domesticates and weedy crops become common, at the beginning of the Early Woodland period. Increasing visibility of EAC crops occurs with different timing in the various subregions: Early Woodland for eastern Kentucky (Gremillion 1993d), eastern Tennessee (Fritz 1993), and the Ozarks rockshelters (Fritz 1993, 1994, 1997); Middle Woodland for central Tennessee (Crites 1978, 1991) and much of the Midwest (Asch and Asch 1985; Wymer 1993); and Late Woodland for the American Bottom of southern Illinois (Johannessen 1984, 1993).

Maize appears first during the Middle Woodland period in widely separated locations but clustered in time between 350 BC and AD 150 (Hart et al. 2007; Simon 2017). These dates lie squarely within the period of Hopewell influence. Hopewell is a regional cultural system that is marked by exchange of exotic trade goods, construction of earthen monuments, mound burial, and a developed mixed economy that included hunted and foraged foods along with EAC crops (Milner and Wills 2013; Smith 1992c). For

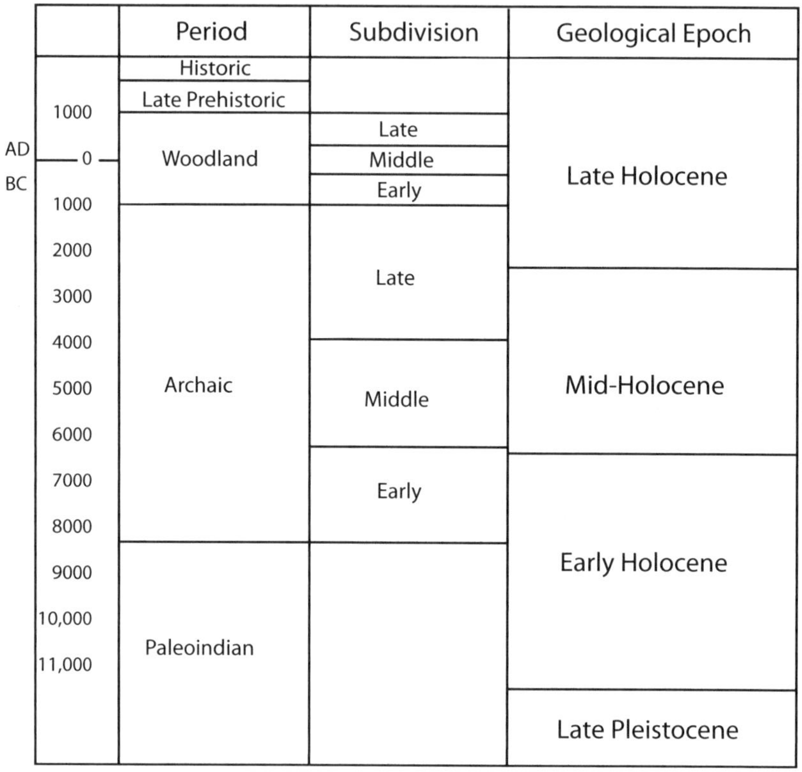

	Period	Subdivision	Geological Epoch
AD 1000	Historic		Late Holocene
	Late Prehistoric		
AD 0	Woodland	Late	
		Middle	
BC 1000		Early	
2000	Archaic	Late	Mid-Holocene
3000			
4000			
5000		Middle	
6000			
7000		Early	Early Holocene
8000			
9000	Paleoindian		
10,000			
11,000			Late Pleistocene

Figure 2.1. Archaeological units for eastern North America. Chronological divisions are approximate.

nearly a millennium following its introduction, however, maize seems to have had little impact; it is only after about AD 750 that it began to achieve a central role as a staple crop (as indicated by carbon isotope concentrations in human bone; Smith and Cowan 2003).

Geographically, archaeological research on the development of food production has been particularly fruitful in several intensively studied river valleys. Figure 2.2 identifies some of these locations, which include the lower Little Tennessee River valley of eastern Tennessee; the Eastern Highland Rim of the Nashville Basin in central Tennessee; the lower Illinois River valley of west-central Illinois; the American Bottom region of southern Illinois; the Ozark uplands of Arkansas and Missouri; west-central Kentucky (which includes the Mammoth Cave area); and the Cumberland Plateau of eastern Kentucky.

Botany and Ecology of EAC Crops

The crops of the EAC were derived from naturally occurring populations of annual weeds (plants adapted to human disturbance that germinate from seed each year). In this they are similar to the ancestors of many major world crops, including some legumes and cereals. However, unlike chickpeas and barley, the EAC crops have not persisted to the present day and are therefore unfamiliar to many archaeologists and the public even in eastern North America. Therefore, a summary of the plants' ecological characteristics and taxonomic affinities is in order. Although taxonomically distinct, all the EAC crops and their free-living relatives share weediness (intolerance of shade and a preference for disturbed habitats) and possession of fruits or seeds that are rich in oil and/or carbohydrates (Table 2.1). In this section, I discuss the four founding members of the EAC (marshelder, goosefoot, squash, and sunflower) and a fifth species (erect knotweed) that emerged after AD 1 in the central Mississippi Valley.

Iva annua L. (marshelder, sumpweed)

Marshelder (Figure 2.3a) is a herbaceous annual member of the sunflower family, Asteraceae, and like sunflower (discussed below), it produces numerous fruits called achenes or cypselae that consist of an oily kernel enclosed within a fibrous pericarp (Gremillion 2004; Wagner and Carrington 2014). The kernels can be harvested efficiently by hand-stripping at maturity (Table 2.2; Smith 1992a). Marshelder is adapted to floodplain soils, where it opportunistically colonizes wet ground exposed by retreating floodwaters, which also serve to disperse its seeds. Marshelder was thus well suited to colonize human-disturbed habitats that are ecologically similar to its natural setting in featuring disturbed soils and ample exposure to sunlight. The domesticated form of marshelder, *Iva annua* L. var. *macrocarpa* Blake, is recognized archaeologically by the large size of its achenes, which exceed the means obtained from free-living populations (Asch and Asch 1978; Yarnell 1972). The earliest clearly domesticated marshelder comes from the Napoleon Hollow site in west-central Illinois (Figure 2.2, Table 2.3; Smith 2006b).

Chenopodium berlandieri Moq. (goosefoot, lambsquarters)

The genus *Chenopodium* includes the currently popular pseudocereal quinoa (*Chenopodium quinoa*), which is native to the Andean region. Goosefoot

Table 2.1 Nutrient Composition of Selected EAC Crops Compared with Hickory Nuts and Maize.

Plant	Energy (kcal/kg)[a]	Macronutrients (% of Total Weight)			Reference	Notes
		Carbohydrate	Fat	Protein		
Domestic sunflower	5700	19	50	23	USDA 2016	Kernels, toasted
Wild sunflower	3813	18	25	16	Schroeder et al. 1974	Weed sunflower (*Helianthus annuus*) achenes
Marshelder	5350	11	45	32	Asch and Asch 1978	Kernels, fresh; Energy content includes fiber
Goosefoot	2729	43	2	17	Asch and Asch 1985:361	*Chenopodium bushianum* fruits, with some perianth parts ("chaff")
Squash	5740	15	49	30	USDA 2016	Pumpkin and squash seed kernels, roasted, without salt
Maize	3650	74	5	3	USDA 2016	Yellow corn
Hickory nuts	6570	18	64	13	USDA 2016	Dried nutmeats

Note: Data from Gremillion (2004)

[a]Edible portion only

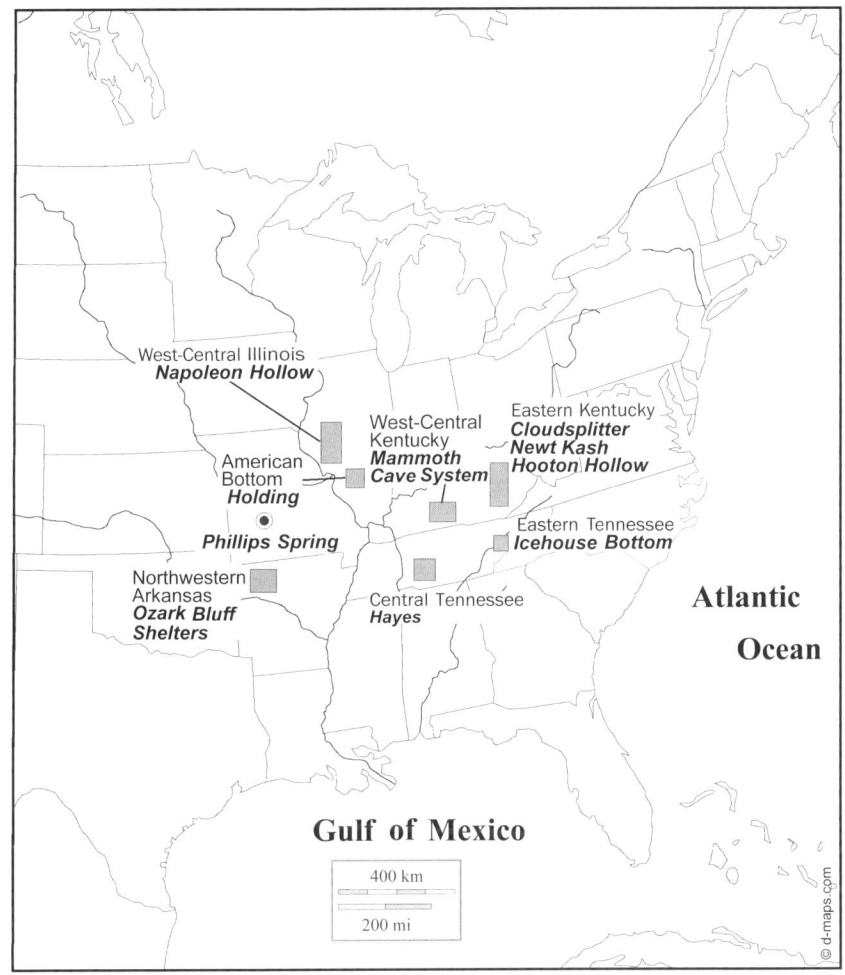

Figure 2.2. Key river valleys, sites, and archaeological research areas relevant to the origins and development of the EAC. Adapted from Smith (1985b:56).

(Figure 2.3b) produces small lenticular seeds that are usually between 1.0 and 1.5 mm in diameter and consist of a starchy interior surrounded by a thin seed coat. The fruits are borne in terminal clusters on branches and axils of the leaves and can be harvested in a manner similar to sumpweed (Gremillion 2014). Also like sumpweed, goosefoot is adapted to floodplains, where it sometimes forms large stands, but has also moved into disturbed upland settings where it is not a strong competitor with other weedy annu-

Figure 2.3. Four domesticates of the EAC: a, *Iva annua* (marshelder); b, *Chenopodium berlandieri* (goosefoot); c, *Helianthus annuus* (sunflower), weed form; d, *Cucurbita pepo* ssp. *ovifera* (*ovifera* squash); d inset, *ovifera* squash seed from Mounded Talus rockshelter in eastern Kentucky (grid size 1 mm); center, assemblage of plant material from Newt Kash rockshelter, eastern Kentucky, with seeds/fruits of the four domesticated species labeled to correspond to whole-plant images. Photo credits: (a) Robert H. Mohlenbrock, hosted by the USDA-NRCS PLANTS Database/USDA SCS. 1989. *Midwest Wetland Flora: Field Office Illustrated Guide to Plant Species.* Midwest National Technical Center, Lincoln, https://plants.usda.gov/core/profile?symbol=IVAN2#; (b) National Park Service; (c) NCTC Image Library, US Fish and Wildlife Service National Digital Library, http://digitalmedia.fws.gov/cdm/singleitem/collection/natdiglib/id/15933/rec/7; (d) Forest and Kim Starr, Starr Environmental (Flickr) CC BY 2.0 (http://creativecommons.org/licenses/), via Wikimedia Commons. https://commons.wikimedia.org/wiki/File%3ACucurbita_pepo_var_ovifera.jpg. Figure inset (center photo) and (d) inset: Photos by Kristen J. Gremillion.

als. Domesticated goosefoot has a reduced or absent outer seed coat, or testa, and a shape that reflects increased volume of the seed interior (Smith 1992a). Its earliest record is from eastern Kentucky, where seeds with thin coats date to the second millennium cal BC (Table 2.3). The name

Table 2.2 Yields and Harvest Rates for Selected EAC Crops.

Plant	Harvest Rate (kg/hr)	Harvest Cost (hr/kg)	Source	Notes
Sunflower (wild)	.17	5.80	Simms (1987)	
Marshelder (wild)	.70	1.42	Smith (1992a)	Average, cleaned achenes, hand-stripped; corrected for indigestible pericarp (shell)
Marshelder (domesticated)	.90	1.11	Smith (1992a)	Based on average 130% increase in seed size
Goosefoot (wild)	1.00	1.00	Smith (1987a)	Estimate from range of 0.7 to 1.1 kg/hr; cleaned, hand-stripped
Erect knotweed (wild)	.90	1.11	Murray and Sheehan (1984)	Harvested by cutting; cleaned achenes

Chenopodium berlandieri ssp. *jonesianum* was coined to honor University of Michigan ethnobotanist Volney Jones (Smith and Funk 1985).

Helianthus annuus L. (sunflower)

Of the four original EAC crops, only sunflower (Figure 2.3c) has persisted as a cultivated plant into the present day. The wild ancestor of the domesticated sunflower (*Helianthus annuus* var. *macrocarpus* [DC] Ckll.) was native to the western United States. From the Great Plains, it spread into the Eastern Woodlands, where it was first brought under domestication. Genetic evidence confirms that these prehistoric populations were the founding stock for all modern cultivars (Blackman et al. 2011; Smith 2014). The earliest known examples of a domesticated sunflower are from the Hayes site in Tennessee and date to cal 2886 BC (Table 2.3). The size of these achenes, with an average of 7.8 mm in length, places them well outside the documented size of wild sunflowers (Crites 1993; Smith 2014).

Cucurbita pepo var. *ovifera* (L.) Alef. (squash, pumpkin)

The native eastern North American *Cucurbita pepo* var. *ovifera* is closely related to the better-known Mexican domesticate, *C. pepo* var. *pepo*, but represents a historically distinct lineage (Figure 2.3d). Genomic research on both ancient and modern material indicates that var. *ovifera* evolved in eastern North America from local free-living populations long before

Table 2.3 Earliest Radiocarbon Dates Obtained Directly from Morphologically Domesticated EAC Crops.

Site	Location	Material Dated	Radiocarbon Years BP	Standard error	Calibrated Age BC (2-σ range)[a]	Calibrated Median (BC)
Phillips Spring	Missouri	Squash/gourd	4440	75	3341–2921	3122
Hayes	Central Tennessee	Sunflower	4265	60	3084–2666	2886
Napoleon Hollow	West-central Illinois	Marshelder	3290	40	1664–1459	1569
Cloudsplitter	Cumberland Plateau, eastern Kentucky	Goosefoot	3450	150	2196–1429	1780
Newt Kash	Cumberland Plateau, eastern Kentucky	Goosefoot	3400	150	2135–1327	1718

Note: Data from Smith (2006b).

[a] Calibrated using OxCal 4.2, online version, https://c14.arch.ox.ac.uk/oxcal/OxCal.html; IntCal13 calibration curve. Sources: Bronk Ramsey (2009); Reimer et al. (2013).

Mesoamerican crops began to enter the region (Decker-Walters et al. 1993; Kistler et al. 2015; Smith et al. 1992). However, in practice it is impossible to distinguish var. *ovifera* material from other fragmentary *C. pepo* rind on purely morphological grounds; instead, material older than about 1,000 years is assumed to be of indigenous origin on the basis of parsimony. Thin rinds and small seeds are characteristic of the weedy forms extant today, such as *C. pepo* var. *texana* (Texas gourd); var. *ovifera* can be distinguished by its relatively thick rind and larger seeds.

Free-living *ovifera* squash is floodplain adapted, occupying an ecological niche similar to that of goosefoot and marshelder. The fruits are dispersed by floodwaters or (during the Pleistocene) by herbivorous megafauna (Kistler et al. 2015; Smith et al. 1992). They contain oil-rich seeds (Table 2.1) and some domesticated varieties also have a fleshy edible mesocarp. Temporal trends in archaeological material indicate selection favoring larger seeds and fruits over time as well as the development of fleshy, thick-shelled varieties (Cowan 1997; Smith 2006b). Modern cultivars of ssp. *ovifera* include acorn, crookneck, and scallop as well as the ornamental gourd.

Cucurbita pepo has played a key role in debates about the origins of agriculture in eastern North America in part because it was once thought to be an exclusively Mesoamerican taxon. Consequently, archaeological seeds and rinds of *Cucurbita* from middle Holocene contexts were thought to overturn the priority of native crops in the Eastern Woodlands (Chomko and Crawford 1978; Watson 1985). Researchers have revised this view in two important ways: first, by recognizing the indigenous status of squash in archaeological contexts in eastern North America; and second, by acknowledging that its mere presence is not an indicator of an agricultural economy. Instead, the small, hard-shelled *Cucurbita* gourds that are represented by the earliest material were probably used as containers or rattles. They were tended for this purpose long before their seeds and rinds show signs of selection under a regime of planting, harvesting, and replanting (Fritz 1999). Selection for larger seeds was underway by 3000 BC (Table 2.3; Smith 2006b). By about 1,000 BC, multiple varieties had been developed to suit different needs for containers, edible seeds, and edible flesh (Cowan 1997).

Polygonum erectum L. (erect knotweed)

Erect knotweed has long been considered a member of the EAC based on its occurrence in seed masses in association with known domesticates (Asch and

Asch 1985; Lopinot 1994). Archaeological collections of erect knotweed often contain some fruits with unusual morphology (large achenes and a smooth pericarp, or fruit coat). Researchers have been reluctant to apply these characteristics as criteria for domestication because they overlap considerably with free-living populations (Dunavan 1993). However, recent work has confirmed the presence of classic morphological markers of domestication in erect knotweed, including larger fruits that germinate rapidly and have thin, smooth coats. Growing experiments and field observations have shown that this domesticated form (*Polygonum erectum* L. ssp. *watsoniae* N.G. Muell.; Mueller 2017) represents a plastic response to cultivation that occurs when selection for fruit heteromorphism is relaxed. Heteromorphism in fruit morphology allows the persistence of different germination strategies, which reduces variance in fitness across time in unpredictable environments. Under cultivation, however, the delayed germination afforded by fruits with thick coats has no selective advantage; instead, planting and dispersal by humans favors consistently rapid germination and growth and thus production of larger numbers of the smooth morph (Mueller 2016a, b). The earliest evidence of domestication based on fruit morphology dates to circa AD 1 (Mueller et al. 2017), making erect knotweed a relatively late addition to the EAC.

Other Crops

Several other native weedy annuals were cultivated in eastern North America, although unlike the five discussed so far they lack the distinctive morphologies that allow researchers to identify domesticated forms in ancient material. These include two cool-season grasses, little barley (*Hordeum pusillum* L.) and maygrass (*Phalaris caroliniana* L. [Crites and Terry 1984; Fritz 2014]). The edible grains of these two grasses were probably important seasonally because, unlike most of the other EAC crops, they mature early in the summer, when plant foods are still in relatively short supply. Giant ragweed and one or more species of amaranth are sometimes found in association with EAC crops and may have been harvested incidentally along with them, or intentionally planted and harvested (Cowan 1985b; Gremillion 1995a).

The EAC in North American Prehistory

Given the difficulty of recognizing minute seeds in archaeological sediments without specialized methods such as flotation to recover them, it is not surprising that the EAC was first discovered in collections of unusually well-preserved plant remains. These collections came from the Ozark "Bluff Dweller Culture" of Arkansas and Missouri, the "Cliff Houses" of eastern Kentucky, and to a lesser extent the Mammoth Cave system of west-central Kentucky. In these cases, plant material normally preserved only through accidental charring during food preparation became highly desiccated in dry, nitrate-rich sediments (Gremillion 1994, 2008). Instead of being consumed by microorganisms, the plants of the EAC that had been stored in caves and rockshelters thousands of years ago survived largely intact long enough to attract the attention of researchers. Although the archaeologists of the early twentieth century had little interest in such material (sometimes even regarding it as "trash": Gremillion 1997), they were unable to ignore such spectacular finds as bundled sheaves of sunflowers or woven bags of seeds. Ultimately it became clear that these archaeobotanical assemblages offered empirical support for a theory of pre-maize food production originally proposed by Ralph Linton (1924).

It was Melvin Gilmore, Curator of Ethnology for the University of Michigan's Museum of Anthropology from 1929 to 1939, who pioneered the paleoethnobotanical study of these desiccated collections. He and his assistant Volney Jones (who later succeeded him as curator) were able to identify many of the botanical remains from the rockshelters of the Ozarks and eastern Kentucky. Their careful analyses revealed that some of the plants represented had unusual morphology, such as the unusually large fruits of marshelder and giant ragweed. Jones (1936) proposed that larger-than-wild-size seeds and fruits from the Newt Kash rockshelter in Kentucky were indicative of domestication. Whereas the giant ragweed size difference proved to be a geographical cline rather than an artifact of domestication (Payne and Jones 1962), and the seemingly large goosefoot turned out to be pokeweed (*Phytolacca americana*), in other cases (marshelder and sunflower), selection under domestication was the best explanation for unusually large fruits and seeds. The domesticates identified by Jones were stratigraphically situated below layers containing maize at Newt Kash, leading him to conclude that they were remnants of a pre-maize agricultural complex (Gremillion 1997; Jones 1936). Furthermore, some of the sites yielding these interesting plant materials contained pits or containers obviously intended for storage of seeds, such as the bag from Marble Bluff in Arkansas (Fritz 1997).

The idea of the EAC was debated following publication of these findings, but received relatively little attention until the flotation "revolution" of the 1970s. In the wake of an article that advocated flotation in *American Antiquity* (Struever 1968), increasing numbers of archaeologists began experimenting with this method for recovery of charred plant material. With the proliferation of highway and dam projects in the Southeast and Midwest during the 1970s and 1980s, large collections of charred plant remains from open (unsheltered) sites became available for analysis. Paleoethnobotanists performed careful quantitative studies that demonstrated the economic importance of EAC crops in the midcontinent during the Woodland period (ca. 1000 BC to AD 800; Asch and Asch 1985; Chapman and Shea 1981; Smith 1992c; Yarnell and Black 1985). Morphometric analyses of seeds and fruits were undertaken in order to develop reasonable phenotypic criteria for differentiating wild and weed populations from domesticated ones (Cowan 1997; Fritz 1984b; Fritz and Smith 1988; Gremillion 1993b, 1993c; Smith 1985b; Yarnell 1971, 1972, 1974, 1978).

Although the EAC gradually lost its detractors, its economic and social significance and its temporal priority over Mesoamerican imports remained contested. The first of these issues was colored by preconceptions regarding the economic potential of native crops. A series of harvesting studies put to rest the contention that these plants were relatively unproductive compared to maize (Cowan 1985b; Smith 1987a, 1992a). Doubts remained, however, that a few scraggly-looking weeds could have provided the economic basis for the mound construction and long-distance trade networks associated with the Middle Woodland Hopewell florescence.

The second issue, that of relative chronology, proved to be more resistant to a solution. Before genomic research was able to demonstrate otherwise, it was common for archaeologists to make the assumption that the presence of squash implied contact with Mesoamerica (and therefore that food production in the East was dependent on cultural diffusion). This association caused a flurry of reassessments of the priority question with the discovery of early cucurbit remains in Late Archaic and even Middle Archaic contexts in eastern North America (Chomko and Crawford 1978; Crites 1987; Kay et al. 1980). It seemed that the EAC had been upstaged by an early introduction of Mexican squash to the East; Mesoamerican "influence" could not be ruled out.

The priority of the EAC was firmly established with a series of taxonomic studies of the species *Cucurbita pepo* that indicated the likely presence of a native weedy lineage in eastern North America (Decker 1988; Decker-Walters et al. 1993; Smith et al. 1992). Early cucurbits no longer posed a challenge to the temporal priority of indigenous crops; instead, they joined the EAC. Maize and beans were still reliable indicators of interaction with Mesoamerican source populations, but these proved without exception to postdate EAC crops. The matter was settled by the advent of direct dating of small quantities of charcoal using accelerator mass spectrometry (AMS; Fritz and Smith 1988; Smith 1992b). Reassessment of chronology based exclusively on directly dated plant remains showed that maize could not be demonstrated to have been present in the East any earlier than circa AD 1 (Fritz 1993). By that date, the EAC had developed into a diversified system that included at least four major crops and several lesser ones. In its early days in the Eastern Woodlands, maize was an addition to a well-established crop complex and not a symptom of revolutionary change in subsistence economy or an invasion from the south.

Ecological and Cultural Processes of Initial Domestication and Dispersal

At the present time, the EAC has become firmly entrenched as a clear example of initial plant domestication that occurred independently of contact with other world centers (Smith 2006b). The ecological context of initial domestication, its chronology, and the roster of species involved are well understood. Topics under debate include the causes of intensification (and the appropriate theoretical framework for explaining this transition), the dietary role and economic importance of native food crops, the incorporation of maize into preexisting systems, and most details of agricultural production (including the ecological structure of garden plots, field location strategies, and the articulation of food production with seasonal mobility).

The Floodplain Weed Theory and the Adaptive Syndrome of Domestication

Current thinking is that initial domestication of EAC crops was underway by 3000 BC. This assessment is based upon earliest dates of domesticated *Cucurbita pepo* from Phillips Spring, Missouri (Figure 2.2; Table 2.3; Smith 2006b). The ecological circumstances surrounding initial domestication of seed crops in eastern North America have been discussed at length by Bruce Smith in the context of the "Floodplain Weed Theory" (hereafter the FWT;

Smith 1992b). In developing this theory, Smith ingeniously drew upon both theoretical and empirical sources of information. His theoretical framework incorporates the ecological approach to human–plant interaction promoted by the biologist and geneticist Edgar Anderson (Anderson 1952, 1956), who viewed the adaptation of certain plants to disturbed habitats as an important pathway to domestication and agriculture (the "dump heap" theory). Smith showed how habitat disturbance promoted the "adaptive syndrome of domestication" described for cereal crops (Harlan et al. 1973). The FWT links the selective pressures found in the seedbed habitat to specific and predictable changes in traits, including most prominently the size of propagules (seeds and fruits) and the control of germination. Competition between seedlings, Smith proposed, selects against smaller seeds (which have less stored energy to fuel rapid growth and reproduction) and those that have mechanisms to inhibit germination (a trait called dormancy that provides some buffer against risk in the wild, but offers no survival advantage once humans take charge of planting).

The dump heap theory coupled with the adaptive syndrome of domestication provided a framework for explaining the emergence of the EAC in the midcontinental region of the United States (Smith 1992b). The process was initiated by human populations that were drawn to settle the newly stabilized land surfaces that formed in river valleys during the mid-Holocene as the hydrologic regime shifted from one of channel incision to aggradation. Such locations afforded ready access to the variety of aquatic habitats then emerging on the landscape, many of which were low-energy wetland systems rich in fish, shellfish, and waterfowl. Repeated occupation of such favored locations maintained the disturbed and open habitats preferred by floodplain weeds. Here, plants such as goosefoot and marshelder flourished. Ongoing soil disturbance and forest opening allowed these weedy annuals to persist near human settlements. Even in the absence of intentional planting, Smith argues, natural selection in this human-modified habitat would have begun moving in the direction of the distinctive morphologies associated with domesticates, such as large seeds and reduced germination dormancy. With consistent planting from harvested seedstock, directional selection was strong enough to move local populations away from the wild or weed state. The floodplain weed theory thus links seed crop domestication to archaeological evidence of increasing residential stability during the mid-Holocene.

Dispersal and Development of the EAC during the Late Archaic Period

Although the four initial EAC crops make their earliest appearances as domesticates in the midcontinental area as defined by Smith (1989), they are usually represented in these cases by small quantities of seeds that suggest a limited economic role during their first few millennia of use. The first domesticates were incorporated into subsistence economies that continued to be based largely on wild animal foods (large and small terrestrial game, fish, shellfish) and plant foods (especially tree nuts such as hickory, acorn, and walnut, and a variety of fruit-producing perennials). The addition of small-scale cultivation does not seem to have required far-reaching changes in social organization, labor allocation, or technology (Smith 2011a; Smith and Cowan 2003).

Recent reevaluation of plant remains from the Riverton site in southwestern Indiana (Smith and Yarnell 2009) supports this view while shedding new light on the evolution of the EAC during the Late Archaic period. Although not a sheltered site, Riverton possesses unusual conditions that have aided preservation of fragile plant materials such as seeds. For this reason, it provides an unusually complete record of subsistence activities, including plant husbandry. Riverton has yielded several morphotypes of goosefoot (including two distinct cultivars and a weed form) along with marshelder, sunflower, and squash. With the exception of goosefoot, only small quantities of seeds were recovered from most deposits at the site, whose archaeobotanical assemblages were dominated by the shells of economically important tree nuts such as hickory and walnut. Thus, Riverton illustrates both the early development of a crop complex centered on goosefoot and the relatively limited role of that complex as a supplement to a stable adaptation based largely on foraging.

It was during this period of initial establishment of food production that the four founder crops were first dispersed beyond their original geographic range in the central United States. Although the precise mechanisms of transmission are unknown, this dispersal relied upon the creation of suitable habitat by humans. Free-living relatives of domesticates such as marshelder and chenopod may well have extended their range in this way. However, in some areas at least, EAC crops were introduced *as domesticates* through interaction between human groups rather than developed from local free-living populations. This mode of dispersal seems to apply to the rockshelter area of eastern

Kentucky on the Cumberland Plateau. Domesticated goosefoot was present on the Cumberland Plateau by 1700 cal BC, but it is not abundant in Late Archaic deposits (Cowan 1985a; Gremillion 1993d, 1994). Domesticated marshelder shows a similar pattern. In eastern Kentucky, EAC seeds with wild- and weed-type morphology are exceedingly scarce. There is no indication that these plants were harvested in the wild prior to their domestication, as they apparently were in west-central Illinois (Asch and Asch 1985). Free-living relatives of EAC crops are at best scarce in eastern Kentucky today (Cowan 1985b). These facts support their introduction to the area as already-domesticated crop plants. Following that introduction, another 700 years would pass before EAC crops were important enough economically to be deposited in large quantities, and in contexts indicating storage.

Even in the absence of morphological markers of domestication, range expansion far outside of the inferred natural distribution of a plant can provide strong evidence for cultivation. The most compelling example of this phenomenon is maygrass, which is confined in the wild to the Coastal Plain of the southeastern United States. Maygrass grains have been found in archaeological contexts well outside the Southeast, as far west as Missouri, Arkansas, and Texas, and as far north as southern Wisconsin (Cowan 1978a; Fritz 2014). This range extension, in addition to the large quantities of grains in archaeological contexts, makes a convincing case for husbandry of maygrass.

Food Production during the Woodland Period

Archaeobotanical evidence shows that nowhere in the Eastern Woodlands did domestication result in a rapid shift to dependence on food crops. It is not until after 1000 BC (the start of the Early Woodland period; Figure 2.1) that EAC crops are increasingly found in large quantities—sometimes even seed masses—and in contexts indicating that they were stored as seedstock or for later consumption outside of the harvest season. Perhaps due to exceptional organic preservation, most of these indications of the growing economic importance of the EAC appear earliest in the Ozarks and the eastern Kentucky rockshelters (Figure 2.2). By about 500 BC (roughly the start of the Middle Woodland period), the impact of seed crop cultivation on human diets and subsistence ecology can be seen within the geographic homeland of initial domestication (Smith 2011a). Between 500 BC and AD

150, food production based on native seed crops became widely established across much of the Eastern Woodlands (Smith and Cowan 2003).

Although the timing of this transition is well documented, we still have much to learn about its causes and consequences, and about the societies that enacted it. With the chronological framework in place, and the ecological context of initial domestication reasonably well understood, research has turned toward questions of process and function. Many of these questions cluster around the details of food production systems and their wider social and ecological impacts. How much did domesticates contribute to human diets, and why did they achieve a more prominent role during the Woodland period? Were crops cultivated in fields, mixed gardens, in uplands, or in floodplains? How much forest was cleared for farming, and how did plant communities change as a result? What impacts did an emphasis on food production have upon social organization, community and household composition, gender roles, status differentiation, and land tenure? Why did the economic importance of native crops vary geographically? And finally, how was maize initially incorporated into the suite of EAC crops?

The EAC and Human Diets

In most cases, archaeobotanists must infer diet (food actually consumed) from a fragmentary record that accrues from the discard and loss of plant parts during processing or consumption. However, for eastern North America an unusual resource is available in the form of desiccated human paleofeces (sometimes called "coprolites," a term usually reserved for mineralized dung; Bryant 1974; Faulkner 1991; Marquardt 1974; Reinhard and Bryant 1992; Stewart 1974). Stable cave environments have preserved human paleofeces in some cases, permitting a more direct assessment of diet from plant tissues that were consumed but not broken down during digestion (Watson and Yarnell 1966; Yarnell 1974). The constituents of paleofeces often include seeds, which are typically enclosed in protective tissues that resist digestion.

The most extensive and best-preserved collections of paleofeces in the Eastern Woodlands come from the Mammoth Cave system of west-central Kentucky (Watson 1974; Watson and Yarnell 1966, 1986). Numerous specimens have been located within both Mammoth Cave and Salts Cave in the passages explored by Early Woodland cavers primarily between approximately 600 and 300 BC. Initial studies of pollen and plant macroremains from the cave material showed that EAC crops formed a substantial proportion of many of the meals consumed. That proportion has been estimated at around 60% (Yarnell 1974), but this generalization masks considerable seasonal and interindividual variability. Seasonal patterns of consumption are indicated by the co-occurrence of fall-maturing and summer-maturing crops (such as goosefoot and maygrass, respectively) in individual specimens (Gremillion and Sobolik 1996; Marquardt 1974; Yarnell 1974). This seasonal pattern suggests one way in which EAC crops enhanced food security for the communities that produced them.

The context of this dietary evidence—deposited during exploration of the deep cave environment—casts some doubt on how representative it is of the full range of foods typically consumed. For example, Smith and Cowan (2003:115) have suggested that the seeds and fruits that are densely packed in many of the paleofecal specimens are residues of "trail mix" consumed primarily by male cavers[2] on expeditions (for further discussion of the "cave food" critique, see Fritz 1993). However, EAC crops are frequent and sometimes abundant constituents of paleofeces recovered from non-cave sites with suitable preservation conditions. Direct evidence that EAC crops were included in meals eaten in an everyday context comes from two sites in eastern Kentucky, the Hooton Hollow and Newt Kash rockshelters (Figure 2.2; Gremillion 1996b). Artifact assemblages and the organization of space at these sites indicate a range of activities including hunting, nut collecting, tool maintenance, textile manufacture, cooking, and storage. The people who occupied them during the Early Woodland period consumed goosefoot, giant ragweed, and other crops. Many of the paleofeces contain large quantities of indigestible plant cuticle, which suggests consumption of less-preferred fallback foods. A similar pattern of seasonal dietary stress was suggested by Cowan (1978b) to explain the presence of material he identified as "inner bark" from the Late Woodland Haystack shelter. Therefore, there is ample evidence of seasonal consumption of EAC crops beyond the deep cave environment.

Explanations for the widespread shift to greater economic importance of EAC crops during the Early and Middle Woodland periods have been tentative. Whereas initial domestication may have been a result of unintentional habitat disturbance and casual cultivation, allocation of increasing amounts of labor to food production implies choosing between alternative uses of time at the individual, household, or community level. Immediate causes that have been favored over the years by scholars of agricultural origins include resource depression (depletion) and risk reduction (see discussion in Winterhalder and

Goland 1997). Resource depression is sometimes a consequence of population growth as increasing numbers of people create a level of demand that approaches the carrying capacity of the local environment. In this situation, cultivated plants may take on an expanded economic role because they respond to intensification (practices that increase harvest yields).

Although promising in theory, explanations of the growing reliance on EAC crops that rely on maximization of economic efficiency have performed rather poorly (Gremillion 2004). Optimization models using the efficiency of energy capture as a currency, such as the diet breadth model of optimal foraging theory, predict that small seeds would be among the least valuable food sources available. Although inexpensive to plant and harvest, they tend to be extremely time-consuming to process into a digestible and palatable form (Table 2.2; Gremillion 2004), resulting in a very low overall return rate. People may have sidestepped this cost somewhat by doing minimal processing for consumption, a practice borne out by the large amounts of waste material from EAC crops found in paleofeces (Gremillion 1996b). The cost of processing may also have been offset by a "back-loaded" storage strategy that defers tasks such as winnowing and grinding until the food is actually needed (Bettinger 2009; Gremillion et al. 2008). Doing so ensures that effort is not wasted in the event that stores go unused and also shifts labor-intensive processing from the busy harvest season when other, more profitable, foods can be sought. These causal pathways are discussed more fully in Chapter 7.

Considering their low net return rates, the EAC crops may have had more potential as lean-season supplements than as year-round staples. Alternative resources such as deer and other mammals, fish and shellfish, and tree nuts yield more calories per unit handling time (Gremillion 2004). However, the small starch- and oil-rich seeds of the EAC have the advantage of being storable (like nuts) and furthermore of being portable. This explanation for the intensification of seed crop cultivation makes sense for eastern Kentucky, where highly valued resources are either scarce (deer) or unpredictable (nut trees that mast, producing crops in some years but not in others; Cowan 1985a; Gardner 1997).

Anthropogenic Landscape Effects and Farming Technology

We know more about the harvesting, processing, and consumption of EAC crops than we do about the wider anthropogenic ecosystems in which these crops were grown. However, combined evidence from sediment cores and archaeological plant remains displays a consistent and strong signal of forest clearing in several well-studied river drainages. For example, in the lower Little Tennessee River valley (Figure 2.2), human communities increasingly relied on uplands and disturbed habitats for fuelwood following the initial appearance of domesticates in the archaeological record (Chapman et al. 1982; Delcourt et al. 1986). The dominance of upland woods in fuel assemblages reflects a decline in the availability of woody vegetation in bottomlands as land was cleared for planting. This interpretation is supported by a sharp increase in ragweed (*Ambrosia* spp.) pollen beginning during the Late Woodland period in the core from Tuskegee Pond, located on a terrace of the Little Tennessee River. A similar pattern of bottomland deforestation characterizes the American Bottom region, where the Middle Woodland increase in EAC seeds accompanied a decline in bottomland wood taxa (Johannessen 1984; Rindos and Johannessen 1991). Thus in the midcontinental heartland of the EAC, alluvial soils were quickly appropriated for farming, whereas uplands remained largely forested until relatively late in prehistory.

In the Appalachian region, the only pollen core that covers the agricultural transition seems to tell a very different story of land use. In the rugged landscape of the Cumberland Plateau, the association of EAC crops with rockshelters (which are usually located on middle and upper slopes) gave rise to the "upland farming hypothesis," which proposed that small plots were cleared on slopes adjacent to shelters (Gremillion and Ison 1993; Ison 1991). Farming of slopes may have had important advantages in a landscape in which stream bottoms were often narrow, choked with canebrakes, and sometimes colder than the surrounding slopes due to temperature inversion. Direct evidence needed to test this hypothesis was lacking until the discovery of Cliff Palace Pond, which yielded a deep sediment core that spanned most of the Holocene (Figure 2.1). Results of palynological analysis from this upland pond showed several interrelated trends in vegetation coincident with the adoption and intensification of food production (Delcourt et al. 1998; Delcourt and Delcourt 1997). After about 1000 BC, pond sediments saw increased deposition of microcharcoal (indicating local burning), pollen of disturbance-loving taxa (possibly including domesticates), and pollen of fire-tolerant trees and shrubs. Delcourt and colleagues (1998) argue from these data that slash-and-burn agriculture in forested uplands created openings in the canopy and altered forest composition to favor fire-tolerant trees.

Composition of wood charcoal assemblages from the Cumberland Plateau indicate that these same tree taxa (including chestnut, oak, and pine) played an important economic role as a source of fuel (as well as food) beginning as early as 7000 cal BC, long before agricultural activity was an ecological factor. These data suggest that the fire effects recorded in the sediments at Cliff Palace Pond may not be representative of the impact of forest clearing across the region (Gremillion 2015).

The farmers of the Woodland period cleared plots in forested bottomlands to a degree that impacted availability of firewood, a phenomenon that is seen clearly in midcontinental drainages that have a continuous record of human occupation throughout the Holocene. However, preference for floodplains was not universal, either due to scarcity of acreage, scheduling conflicts with foraging activities, or risk of flooding. Flooding was certainly one reason why first-order river valleys seem to have been avoided by Woodland farmers east of the Appalachians in favor of second- and third-order streams (Smith 2011a). On the Cumberland Plateau, limitations on availability of level land and alluvial soils would have increased the appeal of colluvial slopes, upland benches, and perhaps even ridgetops as locations for garden plots. The feasibility of upland farming is supported by the presence of relatively fertile soils outside of the floodplain of the Red River (Gremillion et al. 2008; Windingstad 2006; Windingstad et al. 2008).

Gardens of EAC crops were even planted in association with the large hilltop enclosure known as Fort Ancient, located in southwestern Ohio (McLauchlan 2003). The earthen embankments that make up Fort Ancient were constructed during the Hopewell era, a period of monument construction and long-distance trade that coincides with the intensification of farming between circa 550 BC and AD 150 (Milner and Wills 2013; Smith 1992c). Within the earthen enclosure at Fort Ancient are many anthropogenic ponds, probably borrow pits for earthwork construction, two of which were sampled for pollen recovery (McLauchlan 2003). Quantities of pollen from domesticated taxa occurred in unusually high concentrations—too high to be accounted for by the natural growth of weedy relatives of cultivated plants, even on disturbed soils. At Fort Ancient, the large concentrations of nonarboreal pollen indicate that the area within the enclosure, and slightly beyond it, was largely deforested.

The artifactual record attests to an agricultural technology that was relatively simple, but adequate for working friable soils without draft animals or plows. Although digging sticks were used for planting by later maize farmers, the small seeds of EAC crops would have been more easily propagated by broadcasting (Scarry and Yarnell 2011). This strategy may have been less labor-intensive than planting, but required at least some light tillage in order to ensure successful germination and growth (Wagner 2003). Clearing was accomplished through a combination of cutting with stone tools and controlled burning.

Specialized tools for plant food processing are not common in the archaeological record, in part no doubt due to rapid deterioration of organic material. Despite this, unusual environmental contexts in sheltered sites have preserved wooden mortars and pestles as well as storage bags and baskets (Fritz 1984b, 1993; Gremillion 2004; Smith 1985a, 1985b). Processing was minimal in some cases, resulting in consumption of fibrous plant material from seed and fruit coverings. However, increased consumption of small seeds may have been the stimulus for improvements in pottery that permitted heating over a direct flame (Braun 1987; Buikstra et al. 1986). The claim that EAC crops had relatively limited potential for multiple methods of preparation (Smith and Cowan 2003:121) is difficult to assess. For the most part, the ethnographic and ethnohistoric records are silent regarding the EAC crops, which had been abandoned or relegated to a minor role in the maize-based subsistence economies encountered by the first Europeans to arrive in the region (Fritz 2014; Gremillion 2014; Mueller et al. 2017; Wagner and Carrington 2014).

Farming Communities, Land Tenure, and Social Complexity

The human communities associated with the EAC show great variability in size, composition, and complexity. The sites with the earliest records of plant domestication (dating to between 3000 and 1400 BC) are otherwise similar to those that lack such evidence (Smith 2011a). In structure and organization they are typical of the seasonally sedentary foraging way of life followed throughout much of the Eastern Woodlands during the Late Archaic period. Some of them, such as Riverton, probably represent multifamily residential camps occupied during the low-water season of summer and fall (Smith 2011a; Smith and Yarnell 2009). Occupants of these camps dispersed at other times of year for hunting, collecting plant foods, and conducting other specialized activities. These populations shared an effective and stable foraging adaptation focused on a limited set of animal and plant resources, with heavy emphasis on whitetail deer and tree nuts, especially hickory and walnuts. Such foods continued to dominate human diets despite the addition of cultivated plants.

Adoption of low-level food production had no discernible impact upon social organization, although it must have entailed a shift away from communal to exclusive property rights (Bowles and Choi 2013; Crothers 2008). Early farming communities of the Eastern Woodlands were largely egalitarian, with expressions of social difference appearing mainly in the form of burial goods that seem to reflect occupational, gender-based, and other forms of specialization rather than any kind of emerging status hierarchy. Mortuary traditions emphasized corporate kin groups' use rights in the form of communal cemeteries (Smith 2011a). Prominent among grave goods are exotics that moved along regional trade networks that connected coastal and inland areas and moved goods across lines of latitude.

By the time EAC crops had achieved a more central economic role during the Woodland period, it is possible to identify communities that were partly reliant on food production. These usually consisted of small settlements of a few households, located along tributary streams with access to uplands (Smith 1992c). During the Middle Woodland period these communities were tied into a larger network of similar settlements with connections to earthen monuments. Mounds and earthworks are believed to have functioned to facilitate interaction between dispersed communities that held beliefs and perhaps ethnic or kin-based identities in common (Yerkes 2005). Middle Woodland societies also shared a commitment to food production, indicated by the abundance and ubiquity (frequency of occurrence) of crop seeds (Fritz 1993; Simon and Parker 2006; Smith 1992c). Intensification of production may have helped to subsidize long-distance trade, moundbuilding, and status competition in the Hopewell world, although its role is difficult to evaluate archaeologically. Food production was clearly not a prerequisite for monument construction, which predates it by a considerable margin and first occurred outside of the zone of greatest use of the EAC crops, in the Southeast (Saunders et al. 1997).

Regional Variability in the Role of EAC Crops

Although the midcontinent boasts the most ancient evidence of food crops and the earliest intensification of food production, the EAC made inroads into other parts of the Eastern Woodlands as well. Although they do not seem to have achieved the same degree of economic importance in the Southeast, Northeast, and Great Plains (Figure 1.2), EAC crops have been recovered from these regions with increasing frequency as archaeological recovery of plant remains has become standard practice. On the eastern Plains, there is evidence of harvesting of wild or weedy goosefoot as early as 3000 BC, and domesticated sunflower may have been used as early as the Late Archaic period (although these findings remain to be confirmed by direct dating of plant remains). Some Middle Woodland sites indicate cultivation of sunflower and marshelder (Adair 2003; Adair and Drass 2011). Future work may reveal a much greater investment in EAC crops on the Plains than previously thought. To some extent the same may be true of the Northeast, although native cultigens there are generally only sparsely represented (Crawford 2011; Crawford and Smith 2003). Exceptions to this generalization include occasional large deposits of chenopod and knotweed. Sunflower was cultivated in the Northeast mainly for its oil, but it reaches its northern limit in Ontario and suffers some loss of productivity at this latitude. Squash has been securely dated to the mid-Holocene in Maine (Peterson and Sidell 1996). Although its rind is too thin to provide unequivocal evidence of domestication, this find strongly suggests that human agency was involved in its spread. Similar material was recovered from northeastern Pennsylvania that was dated to the fourth millennium BC (Hart and Sidell 1997). EAC crops have been recovered, but generally in small quantities east of and south of the Appalachians. The seemingly limited role played by EAC crops in the Southeast may be related to the region's relatively long growing season and limited usefulness of stored seed crops (Gremillion 2002a). However, expansion of archaeobotanical research over the last decade or so has documented exceptions to this pattern, such as the abundant maygrass from sites in the Piedmont and Coastal Plain of North and South Carolina (Fritz 2014:24–25). Whether or not these examples indicate a much more heavy reliance on EAC crops that was masked by gaps in regional coverage will become more clear as the archaeobotanical dataset expands.

Summary

The shift to greater reliance on EAC crops during the Middle Woodland period was accompanied by an event whose full significance would not be felt for another millennium: the introduction of the Mesoamerican crop maize. The earliest securely dated maize remains from the Eastern Woodlands come from scattered locations, indicating that it was available in the region by 300 BC (Hart et al. 2007). All of these instances consist of small carbonized fragments or even smaller silica bodies that would have remained largely undetectable by archaeologists before specialized methods for their extraction came into common use. These records of early maize are comparable in number to the handful of directly dated seeds that comprise the earliest record of the EAC. Maize seems to have been incorporated into the suite of EAC crops and was presumably grown alongside them, unless the early finds represent trade goods exclusively. In other regions of North America, maize was taken up by human groups accustomed to a foraging way of life, which presents an additional puzzle. Why maize dispersed as it did throughout much of North America, ultimately becoming one of the defining characteristics of many Native subsistence economies, is the subject of the next two chapters.

Notes

1. The term "squash" as used here refers in general to members of the species *Cucurbita pepo* regardless of whether their fruits are fleshy (as culinary usage of the term implies). This species is represented by both Mesoamerican and eastern North American lineages that underwent domestication independently, but it is not always possible to distinguish the two in archaeological material.

2. A sample of 12 radiocarbon-dated paleofeces from Mammoth Cave and Salts Cave was found to contain the hormones testosterone and estradiol (Sobolik et al. 1996). Ratios of these two hormones indicate that the samples were deposited exclusively or primarily by adult males.

3

Origins and Development of Maize-Based Agriculture in the Southwest

This chapter explores the development of maize-based agriculture in the Greater Southwest of North America with particular attention to intraregional variability and its causes. I begin with the introduction of the maize plant, by mechanisms that are still poorly understood, to the hunter-gatherer communities of the Southwest. These groups had developed a sophisticated subsistence ecology that combined hunting, seasonal harvesting of plant foods, and even some quasi-agricultural practices such as broadcast sowing of seeds and controlled burning to increase the density of valued food plants and habitats. As farming communities were established across the Southwest, they developed strategies for resolving the trade-off between adequate rainfall (found at higher elevations) and a sufficiently long growing season (characteristic of the low-elevation drylands). The success of maize agriculture during periods of high rainfall led to demographic expansion during the Pueblo period on the Colorado Plateau; however, growth made populations especially vulnerable to food shortage during droughts.

Culture Areas and Chronology

The Southwest as a culture area (as opposed to a US region) actually incorporates portions of modern-day Mexico and for this reason is sometimes labeled the "Desert Borderlands" or "Greater Southwest." The term "Southwest" as used here encompasses the southern portions of Colorado and Utah, all of Arizona and New Mexico, and the northernmost portions of the Mexican states of Sonora and Chihuahua. Archaeologists generally recognize several distinctive cultural traditions within this area: the Mogollon of

ORIGINS AND DEVELOPMENT OF MAIZE-BASED AGRICULTURE

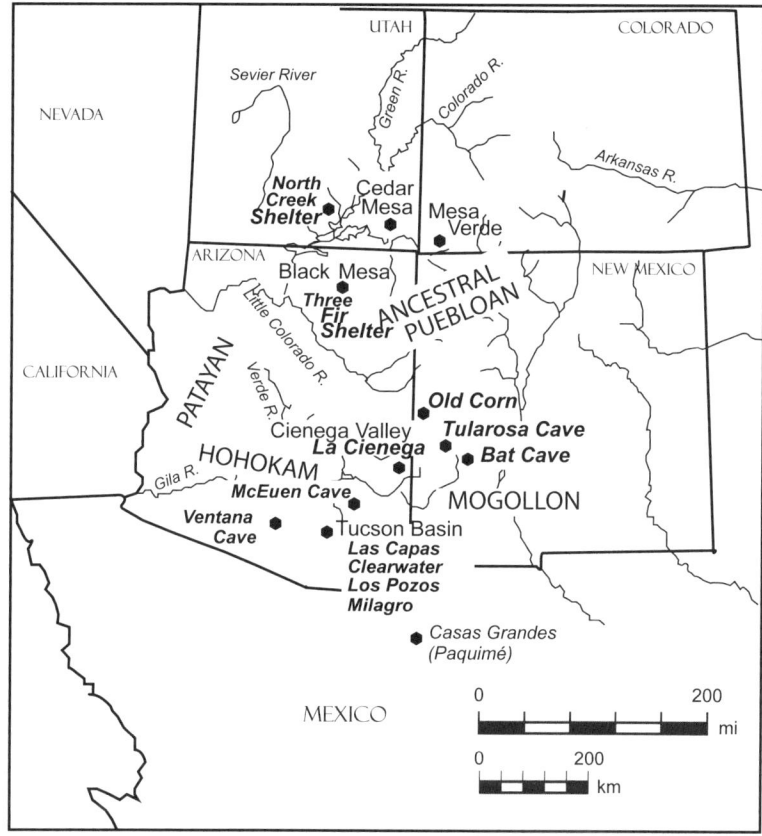

Figure 3.1. Selected sites with early maize in the Greater Southwest. Adapted from Adams and Fish (2011:148). Capital letters designate archaeological subregions.

southern New Mexico and central Arizona; the Hohokam of southeastern Arizona; the Anasazi or Ancestral Puebloan of the Four Corners area (where the four US states of the Southwest meet); and the Patayan in western Arizona (Figure 3.1).

Like any archaeological region, the Southwest has accumulated numerous named archaeological units that are distinctive with respect to time and place. In the interests of clarity and to facilitate comparisons, I will generally ignore these in favor of chronometric dates. I make an exception for the Basketmaker and Pueblo sequence of the Pecos Classification (Figure 3.2), which is a useful framework for the northern Southwest and is still widely

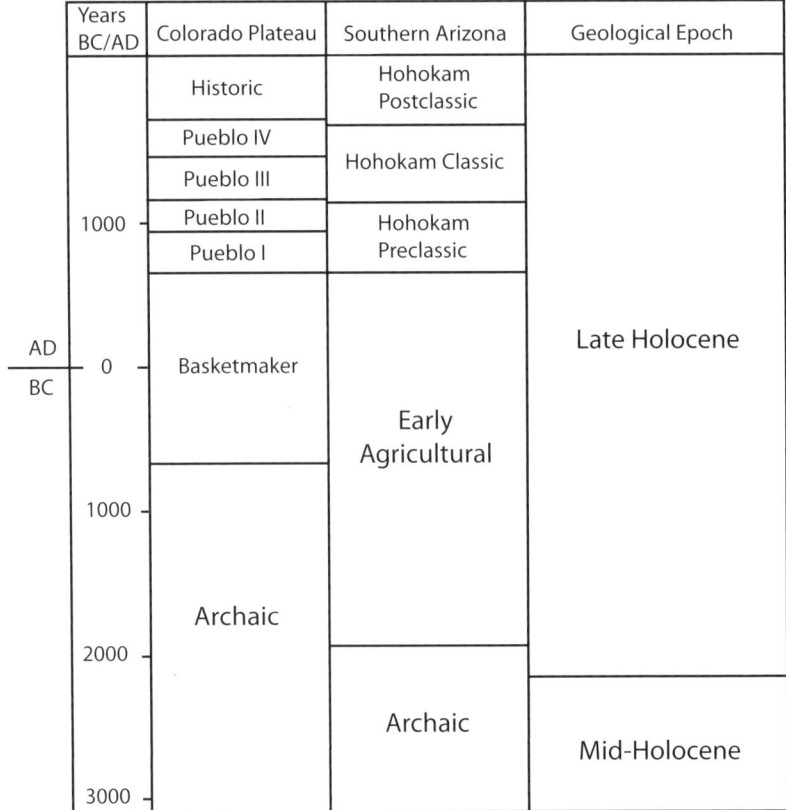

Figure 3.2. Generalized archaeological chronology for the Colorado Plateau and southern Arizona. Source: Milner and Wills (2013).

used. I also include archaeological units for southern Arizona, a particularly important region for recent studies of early maize.

The Southwest is generally characterized as arid or semiarid, but this label masks considerable variability in temperatures, rainfall, and availability of arable land (Roth 2016:10). The dominant landform in the northern portion of the region is the Colorado Plateau, whereas most of the southern Southwest lies within the Basin and Range physiographic province of low-lying deserts separated by mountains (Figure 1.2). Rainfall tends to be higher in the northern Southwest, but scarce surface water and high interannual variation make agriculture a risky venture there (Roth 2016:12). The southern Southwest has relatively low rainfall but a long growing season of

240–250 frost-free days (Roth 2016:90) and major irrigable rivers and, as a result, can be very productive (Mabry 2005). Despite these contrasts, both northern and southern parts of the Southwest share strong altitudinal zonation of vegetation. Strategically located settlements could take advantage of considerable biodiversity within a relatively short horizontal distance.

This chapter focuses on maize-based systems of food production because of their prevalence across North America. However, this focus should not obscure the diverse suite of crops grown by Native peoples in the Southwest, some of which were of local origin. Ford (1981) recognized several distinctive crop complexes of the region, which were revised by Mabry and Doolittle (2008) in light of recent archaeological evidence. In addition to Ford's categories, they propose a possible Early Southwestern Crop Complex based upon locally available weedy annuals and large-seeded perennial grasses. The idea of pre-maize agriculture based on native crops has gained some traction in recent years, but there is no evidence that the plants involved were true domesticates. Those arrived with the Early Mesoamerican Crop Complex, which included maize, squash of Mexican origin (*Cucurbita pepo* var. *pepo*), possibly cotton, common bean, and the bottle gourd, all of which entered the region between 2000 BC and AD 1. There was also a Late Mesoamerican Crop Complex consisting of plants adapted to hot, humid climates (a second variety of cotton, *maiz de ocho* [an eight-rowed maize landrace], jackbean, cushaw squash, butternut squash, and domesticated tobacco) that arrived between 550 and 150 BC. At around the same time, a Late Southwestern Complex developed that featured the tepary bean, agave, little barley, Mexican crucillo, panic grass, and devil's claw.

Introduction and Dispersal of Maize and Other Crops

Maize was first introduced into North America from its source area in tropical Mesoamerica by way of the Southwest. It has long been suspected that maize was descended from the closely related grass, teosinte. Genetic research has confirmed this suspicion, and maize is now placed with teosinte within the same species (*Zea mays*). Studies of microsatellite loci in modern populations of maize indicate that its closest relative is an annual teosinte, *Zea mays* ssp. *parviglumis*. Based on estimated mutation rates, the separation of maize from its shared ancestor with teosinte occurred in the Balsas River drainage of southern Mexico approximately 9,000 years ago

(Matsuoka et al. 2002). These findings are consistent with the earliest known records of maize macroremains, which are from the Guilá Naquitz rockshelter in Oaxaca and have been directly dated to 4300 cal BC (Piperno and Flannery 2001), although older maize remains may yet be discovered. From its place of origin, maize spread both northward and southward, but the archaeological record of its travels is spotty and must be inferred primarily from genetic evidence.

The dispersal of maize throughout North America is closely bound to the history of human migrations and interactions. Maize is highly dependent upon humans for successful reproduction, having lost the ability to disperse its seeds. Therefore, the presence of maize in the archaeological record outside of its place of origin is de facto evidence of plant cultivation and of the transmission of goods and information between human groups. This characteristic of maize removes some of the guesswork involved in assessing the behavioral significance of archaeobotanical maize remains. As we follow maize through the Southwest and into the Eastern Woodlands (see Chapter 4), we can feel confident that it was a crop. Less obvious is the impact maize production had upon subsistence patterns. That impact was highly variable across time and space, until after approximately 1000 AD, when every maize-growing population in North America seems to have entered a new era of dependency on this productive—but nutrient-poor—domesticate.

The spread of maize northward from southern Mexico presented a number of ecological challenges for the plant and its cultivators. While many of these challenges were ameliorated with the development of local landraces throughout its range, adequate moisture during the growing season remains an important limiting factor in the dry climate of the Southwest. Modern maize requires 40–60 cm of rain annually in order to mature, and adequate rainfall at key periods of development (emergence of tassels and ears) is essential. Seasonal rainfall amounts are consequently more influential on yields than annual totals (Fish 2004). The periodicity of rainfall differs across the Southwest, with the western area showing a bimodal pattern of heavy rains in the summer and winter and the east having a single maximum in the late summer. The driest areas tend to be low-elevation deserts that experience both low rainfall (less than 20 cm of annual precipitation) and high evaporation. In contrast, higher elevations are more likely to receive adequate rainfall, but may lack a growing season of sufficient length (120 days for modern maize varieties; Cordell and McBrinn 2012:44–47). Managing this trade-off led to the development of irrigation and other methods of water control and conservation. These intensification and risk-reduction strategies were severely tested late in prehistory when the climate became less stable and droughts more severe.

The earliest dates for archaeological maize cluster around 2000 calibrated radiocarbon years BC (Table 3.1). Dated remains come from a number of sites widely dispersed across the region, indicating diffusion across a broad front rather than a single linear route (Cordell and McBrinn 2012:143). Examples include both rockshelters and open sites and vary in setting from relatively well-watered uplands to low-elevation deserts (Figure 3.1). In these contexts, maize remains are frequently accompanied by those of squash. Another southwestern staple, the common bean, seems to have been introduced somewhat later, by about 500 BC. By AD 1, the bottle gourd (*Lagenaria siceraria*), a globally distributed crop plant originally from Africa (Kistler et al. 2014), was introduced into the Southwest, probably from Mesoamerica (Fritz 2011).

The fact that Mesoamerican crops did not inevitably arrive as a "package" yet appeared more or less simultaneously across a broad geographical area points to diffuse origins of farming rather than a single migration event. However, some archaeologists have made the case that the movement of farming populations into the Southwest was responsible for at least some instances of the early appearance of maize in the archaeological record. A migration of Uto-Aztecan speakers at 2100 BC has been suggested based on analysis of cognate terms related to maize agriculture (Hill et al. 2008; LeBlanc 2008). Support for this scenario has not been forthcoming from archaeology or genomics, however, leading Merrill and colleagues (2009) to conclude that Uto-Aztecan speakers played a crucial role in the diffusion of maize culture but were probably already resident in the Southwest by the time maize appeared on the scene. Association of maize with a migration event would help to explain the relatively rapid establishment of maize-based economies in the Four Corners area by 450 BC following a hiatus in occupation. It would also make sense of the distinctive character of Basketmaker II material culture when compared to its Archaic predecessors (Coltrain et al. 2007). Although the balance of evidence points to a rapid diffusion of maize between groups, migration may have contributed to the spread of farming through the Southwest (Mabry 2005).

Table 3.1. Selected Radiocarbon Dates Obtained Directly on Macrobotanical Remains of Maize, Beans, and Squash from Early Contexts in the Greater Southwest.

Site and Location	Radiocarbon Years BP	Standard Error	Calibrated Age BC (2-σ Range)	Calibrated Median BC	Material Dated
Three Fir Shelter, Colorado Plateau, AZ	3610	170	2470–1540	2260	Maize
Las Capas, Tucson Basin, AZ	3670	40	2200–1940	2050	Maize
Clearwater, Congress St. Locus, Tucson Basin, AZ	3690	40	2200–1960	2080	Maize
	3650	40	2140–1910	2020	
McEuen Cave, Gila Mountains, east-central AZ	3690	50	2270–1940	2080	Maize
Bat Cave, Mogollon Highlands, NM	3740	70	2340–1950	2150	Maize
	3120	70	1600–1130	1390	
	2980	120	1490–910	1200	Squash
	2140	110	2350–1870	2130	Common bean
Old Corn, Colorado Plateau, NM	3600–3800		multiple assays	2260–1980	Maize
Tularosa Cave, Mogollon Highlands, NM	2470	250	1260 BC–AD 20	590	Common bean

Note: Data from Merrill et al. (2009).

Agricultural Strategies and Risk Management

Farmers of the Southwest relied on a variety of strategies to reduce the risks of crop failure in an arid environment. Few locations in the Southwest receive sufficient rainfall to support maize agriculture without efforts to acquire, conserve, and divert water. True dry farming (supported exclusively by rainfall) was relatively rare in the region except in parts of the Four Corners region. Elsewhere, even the earliest maize cultivators relied on accessible groundwater or water control structures to channel runoff into lowland fields (Doolittle and Mabry 2006). In later periods, farmers along the Rio Grande in New Mexico conserved moisture through the use of gravel mulch or rock piles as well as the gridded field systems whose rectangular enclosures trapped moisture and reduced evaporation rates (Dominguez 2002).

Doolittle (2000) offers an extensive review of water control and conservation methods used in the prehistoric Southwest. Canal irrigation was practiced as early as 1300 BC in southern Arizona (Mabry 2005) and appeared on the Colorado Plateau around 1000 BC (Damp et al. 2002; Mabry and Doolittle 2008). Beginning as early as the first century AD, irrigation canals were a key feature of Hohokam land use in southern Arizona. Hohokam farmers created a distinctive subsistence system closely adapted to desert conditions that combined irrigation agriculture with selective utilization of a suite of drought-resistant wild and semicultivated plant species (Bayman 2001; Fish and Fish 1992; Mabry 2005).

Most of the farming practices documented historically in the Southwest can be traced back to the Early Agricultural period. These include irrigation, floodwater farming, water table farming, runoff farming, and farming that relied on rainfall and retained soil moisture (Mabry and Doolittle 2008). Fire was used historically to clear land for planting, and there is some evidence that deliberate burning was practiced in prehistory as well (Adams 2008; Bohrer 1992; Mabry and Doolittle 2008). Other aspects of agricultural technology can be inferred on the basis of artifact assemblages. The implements used in farming were relatively simple; they included stone hoes (recovered from settlements) as well as digging sticks and other tools made of perishable materials (found in some caves in northeastern Arizona; Mabry and Doolittle 2008). Much more common archaeologically are the ground stone implements used to make maize and other seed foods palatable. This technology predated the introduction of maize and other crops to the Southwest, but underwent some changes as it was adapted to the new food source. Over time, the total surface area of manos (hand stones) increased, and metates were deliberately shaped and sometimes set into floor features (Hard et al. 1996). Ethnographically, maize is ground by women and girls. Assuming that this was true in prehistory as well, as maize became a more central food source, women would have spent increasing amounts of time each day grinding.

The farmers of the prehistoric Southwest were able to cultivate maize and other non-native crops by deploying a relatively simple tool kit along with cultural knowledge accumulated across millennia. Despite their successes, interannual and seasonal variation in rainfall stressed crop production systems. In dry years, floodwater farming and irrigation would have provided a buffer against crop failure (Herhahn and Hill 1998), although a multiyear drought can threaten even the most robust water sources. Avoidance of risk (here defined as high variance of an outcome, in this case crop yields, following Marston 2011) may be one reason why southwestern peoples continued to rely to some extent on the drought-adapted native plants of the deserts. Some (such as the inner bark of trees, cactus stems and fruits, and small seeds of annuals) have been labeled "famine foods" (Minnis 1991) because they had undesirable traits (such as bitterness or the need for extensive processing) and were therefore normally ignored except when preferred foods were scarce. Desert plants played an important role as comparatively reliable resources that allowed maize-based subsistence to persist across time in a marginal environment.

Yield estimates suggest that cultivation of agave in northern Mexico would have been an effective buffer against food shortage in maize-based economies when rainfall was particularly low (Anderies et al. 2008). However, agave was not universally relegated to fallback-food status. In the Hohokam area, it was a major cultivated plant that was tended in specialized fields, often on marginal land. This strategy, which began after AD 700, complemented the planting of maize on irrigated alluvial soils (Fish 2004; Minnis 2015).

Early Agricultural Period Communities, 2000 BC–AD 200

It is difficult to generalize about the archaeological character of Early Agricultural period communities, which occur in widely divergent ecological settings and represent historically distinct cultural traditions. The task is rendered even more challenging by the ephemeral nature of some of the sites that have yielded early evidence of maize. However, Early Agricultural societies throughout the Southwest faced similar challenges arising from the patchy distribution of resources across zones that varied greatly in temperature, rainfall, soil properties, and vegetation. Successful subsistence adaptations to these habitats combined seasonal mobility with central-place foraging. Adoption of maize required that these established routines be modified to incorporate the associated planting, tending, harvesting, processing, and storage activities.

The populations that acquired maize and other Mesoamerican crops lived in diverse habitats but shared a foraging pattern that targeted seasonally available resources across a range of upland and lowland habitats. The upland–lowland contrast is an important determinant of plant communities and the animal populations associated with them throughout the region, although species composition and structure vary (Commission for Environmental Cooperation 2011; Huckell 1996). Low elevations supported cacti (such as cholla and prickly pear) and succulents with edible fruits and leaves. Expanses of desert and desert-scrub vegetation were crossed by stream valleys with riparian communities dominated by mesquite or cottonwood and willow. At higher elevations, woody vegetation becomes more common. Forest composition varies, but intermediate elevations usually support piñon-juniper or pine-oak woodlands. Dry mixed forests are replaced at higher elevations by montane coniferous forests. The highest peaks extend above the treeline and support alpine tundra. With multiple biotic zones within reach, most human foragers throughout the Southwest had access to a variety of game, piñon seeds from the piñon-juniper forests, and edible fruits and seeds from the drier lowlands. The latter category might include cactus leaves and fruits, mesquite pods and seeds, and a number of small seed-producing annuals such as goosefoot and amaranths. Grass seeds including ricegrass and dropseed were also collected and consumed (Adams and Fish 2011). These annuals have much in common with plants of the EAC, including broad environmental tolerance and a preference for disturbed habitats.

Husbandry of Native Plants

The abundance of small seeds at some Early Agricultural period sites has caused researchers to ask whether pre-maize subsistence included plant husbandry. Active management of vegetation by foragers is now generally accepted as a widespread practice, but its precise role in the Southwest prior to the Early Agricultural period remains contested. This issue is important because it is relevant to understanding patterns of maize adoption—some have argued that experience with husbandry of native plants served as a preadaptation for agricultural economies (Smith 2005a).

There are sound ecological and economic reasons why people might have intentionally altered vegetation to increase the productivity and predictability of useful plants. Low-intensity fires clear understory plants, induce sprouting, and increase production of seeds and fruits (Delcourt 2008; Hammett 2000). Such practices have been well documented ethnographically and archaeologically in many world regions. In the southern Southwest, the Hohokam used fire to enhance soil fertility and diversity and to clear irrigation ditches of unwanted vegetation (Adams 2008). Other plant husbandry practices documented ethnographically include the broadcast sowing of grasses on floodplains by Colorado River groups. This strategy, which involved only minimal labor input and no care during the growing season, is sometimes referred to as semicultivation (Doebley 1984; Roth and Freeman 2008). In plants like panic grass, devil's claw, and little barley, cultivation and harvesting seem to have led to selection for traits associated with domestication, such as larger seeds and fruits or reduced germination dormancy (Adams 2014; Nabhan and deWet 1984; Nabhan et al. 1981).

These plants were part of what Mabry and Doolittle (2008) refer to as the Late Southwestern Complex, but their status prior to 500 BC or so is not clear. There is direct evidence in the form of seeds and pollen for utilization of grasses, chenopods, and amaranths during the Middle Archaic in southern Arizona. In some cases, they occur in substantial quantities that indicate intensive harvesting (Huckell 1996; Roth and Freeman 2008), and some of the amaranth associated with the San Pedro phase (the latter part of the Early Agricultural period in southern Arizona) may have been domesticated (Mabry 2005). Harvesting and husbandry of small seeds therefore has a long history in the Southwest that may predate the introduction of maize and other crops from Mesoamerica, a pattern widely recognized by researchers (Doolittle and Mabry 2006; Mabry and Doolittle 2008; Minnis 2015; Wills 1988, 1995).

Some argue that husbandry of weedy plants represented a key seasonal strategy for utilization of upland habitats. Sullivan (2015) reports prehispanic pollen spectra from the Grand Canyon area that are dominated by cheno-ams (genera in the Chenopodiaceae, or goosefoot family). These pollen records are associated with in situ ground stone tools and agricultural terraces that are usually interpreted as indicators of maize agriculture. The association of cheno-am pollen with agricultural technology suggests that the husbandry of weedy annuals using controlled low-intensity fires was a crucial component of subsistence (Sullivan 2015; Sullivan and Forste 2014). Fire would have increased the yields of both piñon and weedy annuals, allowing extended or repeated occupations of favored upland locales. Fire management would help to explain the high frequencies of chenopods and amaranths in the archaeobotanical record, although direct evidence of anthropogenic burning is sparse.

The Subsistence Role of Maize

The temporal gap between the initial introduction of maize (1500–2000 BC) and the emergence of economies based largely on food production (AD 500–1000) is sometimes cited as evidence for a lengthy period of use as a dietary supplement rather than a staple (Vierra and Ford 2007; Wills 1995). Opinions differ regarding whether maize was added incrementally in such a way as to be compatible with high mobility or adopted by comparatively sedentary populations already positioned to invest in farming. The former scenario was central to Emil Haury's (1962) model of the agricultural transition, which was inspired in part by the upland provenience of the earliest maize records known at the time. It seemed likely, before the chronological adjustments brought about by direct dating of maize remains, that crops had been adopted initially to supplement food supplies at hunting camps in the mountains (Dean 2005). A version of this model was promoted by Wills, who argued that maize was initially valued because it increased foraging effectiveness by permitting longer residence at hunting camps. According to this scenario, in the Mogollon highlands, maize was especially useful during the spring, when wild resources were relatively unproductive. Cultigens stored in rockshelters also helped to establish control over piñon foraging areas used in the fall (Wills 1995).

The early impact of maize on human subsistence practices varied considerably across the region, representing different levels of labor investment and dietary reliance (Roth 2016). However, where archaeological context is available, the earliest maize in the Southwest is usually associated with subterranean storage pits. This technology may have been introduced along with the plant itself (Roth 2016:35, 136) suggesting that adoption of agriculture was closely linked to the value of maize as a buffer against variation in food supply. This configuration holds whether the sites in question are in the northern or southern portions of the Southwest, open or sheltered, riverine or upland. The Old Corn site, with maize dates that cluster around 2000 BC, was a seasonal foraging camp on the Colorado Plateau with numerous features, including hearths, roasting pits, and storage pits. Las Capas, a deeply buried floodplain site in the Tucson Basin, yielded maize remains dating to 2150 BC and a similar array of features (Roth 2016:65, 104).

After about 1200 BC, maize had become securely established within the subsistence economies of the southern Southwest (Roth 2016:109). Occupations in the Tucson Basin dating to between 1200 BC and AD 200, including the San Pedro phase component at Las Capas, contain abundant maize and structures as well as dense concentrations of storage pits. The farmers who lived there took advantage of the high water table to plant maize, which was further supplied by ditch irrigation (Diehl and Davis 2016). On the Colorado Plateau, Basketmaker II sites (ca. 500 BC–AD 350) typically have bell-shaped pits, pithouses, and slab-lined cists (Huckell 1996). Other evidence of intensive occupation generally postdates the Early Agricultural period. Remains of rodents and birds at sites in southern Arizona indicate that settlement permanence in river floodplains increased only after AD 200 (Dean 2005). The structure of early farming communities led some archaeologists to conclude that the introduction of maize to the Southwest represented more than a minor addition to subsistence, instead acting as a stimulant for significant cultural change (Huckell 1996). However, it is important to keep in mind that these sites are not those of the very first maize cultivators but instead represent cultural patterns that developed after several centuries of use.

Early Agricultural period community organization and settlement continues to be a subject of debate as new evidence comes to light. The long period of consistent, repeated use of the Las Capas site, along with frequent deposition of maize remains, presence of hearths and dwellings, and numerous subterranean features have been cited as evidence of residential permanence associated with food production. Nonetheless, recent analyses of pollen and microorganisms from bell-shaped pits probably used for storage indicate that frequent inundation would have made them useful only for a limited period following harvest (Diehl and Davis 2016). Rather than occupying

Las Capas without interruption, people likely visited the site on numerous occasions across eight centuries, creating the palimpsest for which it is named (Las Capas means "the layers"; Diehl and Davis 2016). In this case at least, early maize cultivators were tethered to their fields for only brief periods and maintained a broad-based diet of foraged foods that required them to relocate seasonally.

Farther south, in northwestern Chihuahua, early maize farmers invested considerable labor in the construction of stone terraces and other rock features at Cerro Juanaqueña. This early agricultural village was occupied during two intervals, 1250–1100 BC and 400–100 BC, that overlap with the San Pedro phase component at Las Capas. Like Las Capas, Cerro Juanaqueña has generated considerable debate regarding the mobility patterns of early maize farmers in the southern Southwest. The extensive rock terracing and other rock features at the site led Hard and Roney (2005:153) to conclude that the site was densely populated from six to nine months per year, or perhaps even year-round. However, others argue instead for a more seasonally restricted pattern of use for agricultural or ritual activities. The widespread presence of maize remains at Cerro Juanaqueña attests to its agricultural function, although its occupants also collected a broad range of wild resources including mesquite, rabbits, cacti, and grasses (Roth 2016:126). Interpretations of site function are also complicated by the apparent defensive nature of its hilltop location. For the time being, Cerro Juanaqueña is distinctive among early agricultural sites in the Greater Southwest.

Whereas some aspects of site organization suggest a significant investment of effort in the cultivation and storage of maize during the Early Agricultural period, direct evidence of the role of maize in the diet is more difficult to interpret. Some of this evidence comes from the chemical composition of human bone (Ambrose 1987; Chisholm et al. 1982; Klein 2013). Carbon incorporated into bone tissue reflects the isotopic composition of its source. Plants that use the C_4 metabolic pathway, an adaptation to low-moisture habitats, are enriched in ^{13}C, a stable isotope of carbon. Consumption of these plants, which includes maize, creates an elevated ^{13}C signature in body tissues, including bones and teeth (Figure 3.3). Elevated levels of ^{13}C also result from the consumption of other species of C_4 plants and the animals that feed on them. Consequently, carbon isotope values alone are ambiguous indicators of maize consumption in the region and must be considered alongside other relevant sources of data (Coltrain et al. 2007).

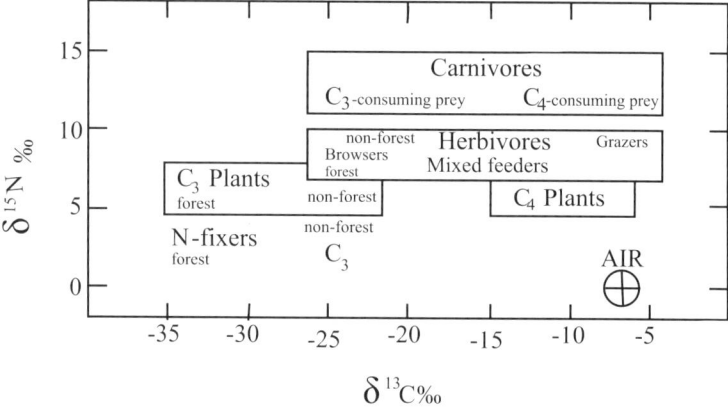

Figure 3.3. Variation in carbon and nitrogen isotopes across trophic levels. Adapted from Hedman (2006).

Studies of bone chemistry in the Four Corners area have been used to make a case for maize dependence among Basketmaker II populations by 450 BC (Coltrain et al. 2007). These authors argue that high ^{13}C values in this case are probably caused by maize rather than other foods, based on corroborative evidence from paleofeces whose contents were dominated by maize and piñon macroremains. Temporal sequences from Mesa Verde, Black Mesa, and Cedar Mesa (all in the Four Corners area) indicate ^{13}C values consistent with heavy maize consumption from 250 BC onward (Chisholm and Matson 1994; Hard et al. 1996). In contrast, in parts of southern New Mexico convincing evidence of maize dependence does not predate AD 50, following a millennium of persistence as a minor crop. A gap of similar length between initial use and establishment as a staple characterizes southeastern Arizona (Mabry 2005). Clearly there was much variation within the Southwest in the timing of this subsistence transition, which followed a long period of limited use in some localities but in others seems to have followed quickly after initial acceptance (Hard et al. 1996).

Assemblages of plant macroremains from sites across the Southwest (Fish 2004) indicate that in most locales, the ubiquity of maize remains increases rapidly following its initial introduction. Ubiquity is a measure of frequency of occurrence in which all instances have the same value regardless of the total quantity of material represented in each case. Therefore, high ubiquity values can be taken to mean that maize quickly became widespread within the communities that used it, but these values do not address differences between households or samples in the quantity of maize produced, processed, and stored. This finding seems at first glance to be somewhat at odds with the carbon isotope evidence, but it is entirely possible that maize production became normative almost immediately without eliminating variation in the amount consumed seasonally or between individuals.

Anthropogenic Impacts

Although human impacts on vegetation, soils, and other landscape elements did not originate with maize agriculture, in many cases they intensified as people cleared fields to plant crops. Husbandry of native plants seems to have emphasized enhancing the productivity or persistence of existing plant communities, whereas maize farming required the establishment of new resource patches (resource clumps separated by nonproductive zones). In areas where rainfed agriculture was not feasible, even the earliest maize farmers had to invest in landscape modifications such as dams, ditches and canals, or terracing to maintain productive fields. This need to supplement rainfall with other water sources, such as groundwater, flooding, or directed runoff, required some combination of strategic field location with construction that modified soil characteristics and microtopography in persistent ways (Adams 2008; Doolittle and Mabry 2006; Spielmann 2008).

The impact of the earliest farmers on local landscapes was relatively light in comparison with that of subsequent periods. Indicators of anthropogenic disturbance such as bones of some rodents and perching birds do not increase significantly in the Tucson Basin and northern Chihuahua during the Early Agricultural period compared to Middle Archaic foraging populations that preceded them. Despite the presence of fields, irrigation canals, storage pits, and structures, early farmers apparently did not modify habitats in ways that significantly diminished populations of animals, such as cottontails and mule deer, that prefer relatively dense vegetation cover (Gilman et al. 2013). These findings indicate that removal of woody vegetation to create agricultural plots was initially limited in intensity and/or extent. There is no evidence that early agricultural settlements attracted commensal species or degraded the habitats suitable for game animals.

Diversification, Intensification, and Social Change, AD 200–1400

After about AD 200, there is increasing evidence for maize-based farming systems and associated impacts on soils, vegetation, and fauna. Pottery vessels were in common use to store and prepare maize and other plant foods, and introduction of the bow increased hunting success in wooded habitats (Cordell and McBrinn 2012:156). These technological innovations contributed to food security and hunting efficiency while supporting longer residence in settlements. A general trend to greater residential permanence was expressed architecturally by a shift from semisubterranean pit dwellings to aboveground structures of adobe or stone, beginning in the Pueblo I period (ca. AD 700) in the northern Southwest. These pueblos (or houses situated around courtyards in the Hohokam area) typically have multiple large-capacity storage pits and often include identifiable public areas and ceremonial structures.

Although some archaeologists have proposed that agricultural intensification created a need for more housing and structures for storage and food preparation (Gilman 1987), the timing of events in some cases is inconsistent with this scenario. A comparison of four different regions within the Southwest between approximately AD 200 and 800 indicates that population growth, agricultural intensification, and depletion of game were all underway before the transition to aboveground housing (Diehl 2012). Rates of dental caries (a frequent consequence of diets high in complex carbohydrates) from more than 100 human skeletons (Schollmeyer and Turner 2004) remained consistently high across the Pithouse-to-Pueblo transition in southwestern Colorado. It seems that although maize agriculture required greater residential stability than the hunting and collecting that preceded it, there is no single trajectory that describes the causal relationship between farming and settlement permanence across the Southwest.

By AD 900, economies at least partly reliant on food production were widespread throughout the Southwest. This expansion may have fueled further agricultural intensification and contributed to the emergence of more complex socioeconomic systems that included extensive trade networks, more formalized decision-making structures, and emerging social inequality. Abundant maize harvests yielded surplus foodstuffs that could be stored, used to feed artisans when they were not doing agricultural labor, or exchanged with other communities. Periods of peak agricultural production therefore offered opportunities to expand social networks and engage in specialized production of exportable goods. They also supported population growth, making human societies increasingly vulnerable to the risk of catastrophic food shortage when environmental conditions deteriorated.

Regionally distinctive cultural traditions become more evident in the archaeological record after AD 200. Farmers in the north (Ancestral Puebloans), the uplands of east-central Arizona and southern New Mexico (Mogollon), and the southern deserts (Hohokam) developed crop complexes and agricultural technologies tailored to each region's environmental opportunities and constraints. Because of the difficulty of generalizing across the Southwest, I discuss the patterns of agricultural intensification, settlement, and socioeconomic organization for each regional tradition separately. For the Ancestral Puebloan area, I discuss Chaco Canyon and Mesa Verde as different solutions to the challenge of maintaining a complex, maize-based socioeconomic system in a region of variable rainfall. In the Mogollon area, Classic Mimbres sites illustrate the environmental impacts of food production in a mountainous zone with rich alluvial soils. The site of Paquimé (Casas Grandes) is notable for its terraced fields and evidence for communal agricultural labor. Finally, the Hohokam area is famous for its extensive network of irrigation canals and ancient tradition of drylands agriculture based on maize and desert-adapted plants. I return to the Southwest as a whole for a discussion of demographic expansion of agriculturally based populations.

Ancestral Puebloans: Regional Integration and Environmental Challenges

Compared to the southern deserts, the Colorado Plateau generally receives more rain, but growing seasons are relatively short and summer rainfall is varied and unpredictable (Adams and Fish 2011:161). The rivers here are not generally suitable for canal irrigation, but farmers used many strategies to direct water to their crops, including the well-known *akchin* fields, which were placed on alluvial fans (Ford and Swentzell 2015). Although there is much variation between sites, most communities of the Ancestral Puebloan region were at least somewhat dependent on maize production by the latter part of the Basketmaker II period (500 BC to AD 400; Adams and Fish 2011:163). Based on the analysis of desiccated human paleofeces, maize was complemented by a variety of lesser crops, wild plants, and game (Minnis 1989). These data indicate that maize was one of several core elements of Puebloan diet that remained consistent across the Pithouse-to-Pueblo transition and until about AD 1300. Other core elements of the plant diet were weedy taxa that took advantage of the open habitats created for cultivation of crops.

AGRICULTURE AND SETTLEMENT IN THE NORTHERN SOUTHWEST. In the northern Southwest, the transition from living in pithouses to mostly aboveground pueblo architecture dates to around AD 700, and the Pueblo I period began between 700 and 1000 AD. The earliest pueblos in this region consist of linear blocks of rooms, usually arranged in suites, each of which incorporated living and storage space. Research by the Dolores Archaeological Program in southwest Colorado shows that large Pueblo I settlements (AD 700–900) were relatively short-lived, perhaps indicating the failure of social integrating mechanisms or the challenge of sustainable resource use (Kohler and Matthews 1988; Schachner 2001; Wilshusen and Ortman 1999).

Anthropogenic impacts associated with agriculture left traces in the record of charcoal deposition, soil amendments, and water control structures. There is evidence of increased burning in the Mesa Verde region between AD 500 and AD 1250. The charcoal record shows a correlation between human population expansion and fire frequency, suggesting active land management by agricultural populations (Herring et al. 2014). Other types of management enhanced the availability of nutrients and water for crops. Water control and conservation efforts provide a hedge against unpredictable variation in rainfall, even in the parts of the Colorado Plateau that are suitable for dry farming. To that end, Ancestral Puebloans of the northern Rio Grande area created sophisticated field systems and enriched soil fertility with floodwaters and the addition of refuse and ashes to fields (Adams 2008:126). To conserve limited moisture, Pueblo IV period farmers on the Colorado Plateau constructed pebble-mulched gardens whose rectangular enclosures trapped moisture and reduced evaporation rates (Dominguez 2002).

THE CHACOAN SYSTEM. The Chacoan system, centered on Chaco Canyon in northwestern New Mexico, is recognizable by its sophisticated architecture and network of roads, although the nature of the "system" is a topic of much debate. In Chaco Canyon itself, and at a few other sites, people constructed massive Great Houses, elaborate multistory stone masonry structures. Elsewhere, across much of the northern Southwest, people built smaller versions, in most cases apparently emulating the Chaco style (Van Dyke 1999). Large quantities of goods—including ceramics, chipped stone, turquoise, and wood—moved into Chaco Canyon from across the northern

Southwest. Other imports came from the south, including copper bells, macaws, and cacao (Crown and Hurst 2009).

The canyon itself is quite dry today, but it still would have been highly productive for Puebloan agriculture. Vivian (1990) showed that large flows of water would have come into the canyon's north side and were directed to fields through a sophisticated water control system. The location of the three earliest Great Houses near the confluence of drainages supports the argument that control of water and/or agriculture was an important component of the rise of Chaco. Recent work by Dorshow (2012), which takes into account the broader landscape, concludes that the agricultural productivity in and around Chaco Canyon was even greater than had been estimated earlier. Agricultural surpluses may have been used to support craft workers whose products could be exchanged for food during less productive times. Turquoise may have played this role in the Chacoan system (Kantner 2010).

Strontium isotope analyses of archaeological material document the importation of maize from as far as 80 km away from the canyon itself (Benson et al. 2003). More recently, isotope evidence has been collected that demonstrates that most of the deer, rabbits, and prairie dogs consumed at the Great House of Pueblo Bonito came from over 40 km away (Grimstead et al. 2016). While it is possible that the communities living in Chaco Canyon were vulnerable to food shortage and relied on long-distance transport to remain viable in the face of adverse environmental conditions, food imports may have been a form of tribute.

THE MESA VERDE REGION. To the north and west of Chaco Canyon lies the Mesa Verde region (Figure 3.1). The region is of particular interest to archaeologists because of its complex population history. Following centuries of overall demographic growth (despite local fluctuations) beginning around AD 600, the Mesa Verde region was rapidly depopulated after AD 1200. Researchers have identified two major cycles of population aggregation followed by depopulation in the northern San Juan Basin, from AD 600 to 920 and between AD 920 and 1280 (Kohler, Varien et al. 2008). Demographic reconstructions within the study area of the Village Ecodynamics Project (Kohler 2010; Kohler et al. 2012) have shown that peak populations coincided with in-migration from nearby locations that were experiencing resource scarcity due to drought. In order to better understand how changing resource abundance would have influenced the decision to abandon the region, an agent-based model was constructed that set virtual agents (representing households) the task of optimizing settlement location to minimize the costs of meeting basic requirements for maize, protein, water, and fuel (Kohler 2010; Kohler et al. 2012). Multiple iterations of the model were run in order to compare the decisions of virtual agents to the archaeological record.

Comparisons of modeled results to the record of site size and location yielded a number of important insights into how maize production articulated with other elements of the Mesa Verde subsistence economy. An early version of the model indicated rapid depletion of deer populations and produced human population estimates that were considerably smaller than those observed in the archaeological record. A revised version of the model that incorporated protein obtained from turkeys (*Meleagris gallopavo* L.) offered a better fit to data, especially after AD 1020 (Kohler et al. 2012). This project illustrates how agriculture can support population growth even when yields are highly variable, but it also demonstrates the impact of other limiting factors on community viability, such as the availability of game. To address the resulting shortage of protein, maize was not only consumed by farmers but also diverted to provision domestic turkeys.

Recent investigations in the Mesa Verde region have played a key role in documenting the domestication of the turkey in the Southwest, the only known example of animal domestication in North America (Lipe et al. 2017). Most of the evidence for this process comes from the Mesa Verde region, particularly during the Pueblo I–IV periods (Figure 3.2). Mitochondrial DNA haplogroups from archaeological material show a predominance of affinities with eastern North American and Rio Grande subspecies rather than with the local wild *M. gallopavo merriami* currently resident in the Southwest (in which haplogroup H2 occurs) (Newbold et al. 2012). However, recent studies of the two main haplogroups represented archaeologically reveal a more complex history. Most of the archaeological turkey remains tested are affiliated with the eastern and Rio Grande subspecies, but a small proportion (15%) represent occasional capture and integration of locally available wild birds into the flock (Lipe et al. 2017). The relatively high degree of genetic uniformity in the archaeological material indicates ongoing evolution of a rather small and isolated population (Munro 2011).

The earliest turkey remains in archaeological context come from the North Creek shelter in southern Utah and date to between 7000 and 7500 BC. The next most recent date is 300 BC from Tularosa Cave (Figure 3.1;

Newbold et al. 2012). The domesticate status of the turkey is usually inferred from archaeological contexts that indicate penning, hatched eggs, turkey droppings, and other signs of husbandry. There is good evidence that people were keeping turkeys by sometime between AD 500 and 850, and this evidence becomes much more abundant after AD 850. By the Pueblo I period, turkey-keeping was a widespread practice (Munro 2011).

In addition to providing food, turkeys were sources of feathers for arrow fletching, decoration of textiles, ritual performance, and other non-food uses (Lipe et al. 2017). However, the protein provided by turkeys may have been a key element of the adjustment of Mesa Verde area populations to drought conditions during the Pueblo II and III periods (Kohler 2010). Supporting this interpretation is evidence that some maize was diverted from human use to become turkey fodder (Kuckelman 2010). A major period of intensification of turkey production took place during Pueblo III, just before the region was abandoned (Munro 2011). Turkeys may have been initially domesticated to provide feathers or sacrifices for ceremonies, but they had an important dietary role to play in times of subsistence stress.

Mogollon: Classic Mimbres and Paquimé

The Mogollon culture area (Figure 3.1) is dominated by basin-and-range topography of alternating mountains and valleys. Contrasts in elevation create a variety of habitats, ranging from the low-lying deserts and grasslands to the coniferous forests of higher elevations. In many parts of the Mogollon area arable land was patchy and scattered, but in southwestern New Mexico, the broad valley of the Mimbres River offered productive soils for farming (Broughton et al. 2010). The Mogollon–Mimbres sequence was defined by Anyon and colleagues (1981) and reviewed by Hegmon (2002) with recent refinements of chronology by Anyon and colleagues (2017).

During the Early Pithouse period (AD 100–550), many Mogollon sites were situated on bluffs and ridges in relatively inaccessible locations (Cordell and McBrinn 2012:172). Sites of this period typically yield large quantities of maize, beans, and squash along with a variety of weedy plants including sunflower, amaranth, and goosefoot (Huckell and Toll 2004:64). Larger and more numerous settlements appear during the Late Pithouse period, from AD 550 to 900, along with public ceremonial structures and indications of expanding trade networks (Cordell and McBrinn 2012:182). Diehl (1996) argued for a major increase in maize dependence coincident with the Pithouse-to-Pueblo transition in the Upland Mogollon area. His claim rests on the higher frequency of maize remains and highly efficient ground stone processing tools at pueblo sites as compared to pithouse sites. However, close examination of taphonomic factors affecting archaeobotanical assemblages implicates differences in the deposition of plant macroremains between the two site types rather than variation in diet (Rocek 1995).

Minnis (1978, 1985) documented increases in several indicators of disturbance in the Mimbres area of southwestern New Mexico, including a drop-off in representation of riparian woods in charcoal assemblages and an increase in weed seeds during the Classic Mimbres period (AD 1000–1130). This pattern probably reflects increased clearance of riverine habitats for agricultural use, and coincides with population growth and expansion into marginal settings (Minnis 1978; 1985:64). Both of these datasets point to land clearance and soil disturbance in riverine settings, conclusions supported by subsequent research (Schollmeyer 2005, 2011). At the same time, the first evidence appears in the Mimbres area of rock borders and check dams for water control (Minnis 1985:61). Population growth seems to have also placed pressure on animal populations, leading to depletion of large-bodied (and thus high-value) animal prey, probably well before the Classic period (Cannon 2000, 2003; Schollmeyer 2011). Resource stress coupled with an unusually severe dry period contributed to depopulation of much of the Mimbres River valley after AD 1150, although people reorganized and continued to live in parts of the region (Hegmon et al. 1998; Nelson et al. 2010) and continued to use the same set of resources (Nelson 1999). Settlements near rich alluvial soils become more common during this period, and expansion into less productive habitats such as terraces suggests competition for arable land (Huckell and Toll 2004:64; Minnis 1985).

CASAS GRANDES (PAQUIMÉ). Farther south, control of agricultural production by elites developed in the Casas Grandes polity of northern Chihuahua during the El Medio period (AD 1150–1450; Figure 3.1; Minnis et al. 2006). The landscape around the site of Paquimé features numerous dry-stone rock alignments known as *trincheras* in northwestern Mexico. These *trincheras* are variable in morphology and function, but all serve to improve conditions for farming by controlling water, improving soil quality, or creating flat terraces. The largest *trinchera* fields in the Casas Grandes area are associated, not with densely populated settlements, but with small, nonresidential sites believed to have an administrative or ceremonial function.

These large fields have been interpreted as analogous to the "cacique fields" of ethnohistoric accounts, which were sometimes farmed communally. The products may have been used to support leaders and to fund feasting and public displays important in forming and maintaining alliances (Minnis et al. 2006).

Hohokam

The farmers of the Hohokam area were able to take advantage of the long growing season of southern Arizona by developing and managing networks of irrigation canals. Such practices were essential in the Sonoran desert landscape with its low rainfall (less than 25 cm annually) and high temperatures at low elevations (Bayman 2001). The well-known irrigation canals of the Phoenix Basin were initiated by AD 600, although they became much more extensive after AD 1000 (Adams and Fish 2011:164–165). The alluvial setting of these irrigated fields ensured ongoing renewal of soil nutrients from flooding (Fish and Fish 1992). Other less labor-intensive techniques were also in use by this time, including stone check dams, rock piles, wells, and reservoirs (Bayman 2001:275). Water control efforts assisted the cultivation of maize, bottle gourd, cotton, and beans.

Hohokam farmers also grew a suite of native desert-adapted plants after AD 500, including little barley grass, tobacco, tepary bean, agave, and Mexican crucillo (Mabry and Doolittle 2008:56). Specialized agave-roasting pits and pebble-mulched fields document the economic importance of agave in the Hohokam area, where it was cultivated on marginal land from AD 700 onwards (Fish 2004). High crop diversity and the inclusion of drought-resistant plants buffered Hohokam communities from unpredictable variation in the yields obtained from maize. The precarious nature of desert agriculture also ensured a continuing subsistence role for locally adapted wild plants such as mesquite and cacti (Bayman 2001).

Hohokam impacts on the land were significant, but in many cases constitute the kind of intermediate-level disturbance that can have positive effects on biodiversity. Bird and rodent remains postdating AD 200 indicate that such disturbances favored species more tolerant of "open" habitats that proliferate near human settlements (Dean 2005). Hohokam people used fire to manage vegetation and enhance soil fertility and biodiversity (Bohrer 1992), much as the Tohono O'Odham and Akimel O'Odham do today (Nabhan et al. 1982). Irrigation created "second gardens" of wild and semiwild plants that thrived on excess runoff (Adams 2008; Fish and Fish 1992). Practices associated with agriculture thus had ramifying effects through local ecosystems, often in ways that improved the abundance or reliability of valued (nonagricultural) resources.

Reliance on canal irrigation can set the stage for centralized administrative management of agricultural activities. However, Hohokam farmers organized agricultural tasks by forming cooperatives across communities (Nelson et al. 2010). Effective management of irrigation infrastructure permitted a highly productive and resilient agricultural system to develop along the Salt and Gila Rivers. Starting around AD 800, there is evidence of a regional interaction sphere involving shared religious beliefs expressed in a ceremonial ballgame. Activities at the ball courts found at some Hohokam sites drew participants from a variety of environmental settings and provided a venue for exchange of goods (Nelson et al. 2010). The collapse of the regional ball court network after AD 1070 removed an important mechanism for limiting risk and disrupted the social networks that were central to irrigation management. In the Hohokam case, cooperative irrigation management sustained a highly resilient agricultural system but was itself vulnerable to perturbations of the social networks upon which it depended (Nelson et al. 2012; Nelson et al. 2010).

Maize and Demographic Expansion

The Neolithic Demographic Transition (NDT) is a trend associated with the transition to agricultural subsistence worldwide. It takes the form of increased birthrates combined with stable mortality rates (Kohler, Glaude, et al. 2008). Maize agriculture does seem to have contributed to such a demographic transition in the Southwest, but not until the first millennium AD (Kohler and Reese 2014). Birthrates reflect an increase in female fertility consequent on changes in women's activity levels and diet with agricultural subsistence. However, rather than lasting several hundred years as it did in other regions (such as the Near East), the NDT in the Southwest was even longer in duration and spatially variable. A proxy for birthrate derived from skeletal remains, the juvenility index (number of individuals 5–19 years of age divided by all individuals over 5 years), does not rise sharply in the Southwest until more than two millennia following the introduction to maize. Analysis of the data by subregion highlights differences between northern and southern portions of the Southwest. The NDT in the southern Southwest was long and gradual, due to the difficulty of expanding irrigation systems and the social inequality that may have limited reproduction among less privileged segments of the population. Waterborne pathogens may also have elevated child and infant mortality in irrigation-dependent areas. On the Colorado Plateau to the north, where maize cultivation could be practiced without extensive irrigation, population growth commenced sooner with respect to initial introduction of maize (on the order of hundreds of years at most, compared to a thousand years for the southern Southwest). After AD 1000 a decline in birthrate can be seen throughout the Southwest, although the timing varies by subregion. In many cases these declines occurred in response to periods of drought or other adverse effects on growing conditions (Kohler, Glaude, et al. 2008; Kohler and Reese 2014).

Population growth during periods when conditions allowed large maize harvests sometimes resulted in depopulation when conditions worsened again. Particularly in locations that were marginal for maize agriculture, the NDT set the stage for subsequent food stress and outmigration. On the southern Colorado Plateau, the effects of drought on human populations have been documented in detail using tree-ring-based estimates of the Palmer Drought Severity Index (PDSI) in conjunction with tree-cutting dates as a proxy for human population (Benson and Berry 2009; Benson et al. 2007). Significant climate oscillations occurred there between AD 860 and 1600 and included six megadroughts (defined as a period exceeding 20 years with PDSI values of less than or equal to –1 for at least 60% of dry period years, and at least two additional droughts at that level that last three or more consecutive years). The megadroughts of the mid-twelfth and late-thirteenth centuries AD were one of multiple factors that resulted in massive depopulation of the Four Corners region. Although not dependent upon canal irrigation as in the Sonoran Desert, maize farming on the Colorado Plateau was still a risky strategy. Floodplain and rainfed farming were highly sensitive to drought; human populations that had grown during favorable wet periods between AD 1045 and 1129 and AD 1193 and 1269 were not sustainable under the dry periods that followed (Benson and Berry 2009). It seems unlikely that this boom-and-bust pattern would have developed without the potentially high yields afforded by maize.

Summary

Prior to 2000 BC, the deserts and uplands of the Greater Southwest were occupied by groups that were sustained by hunting and the collection of plants long adapted to the dry climate. Foraging strategies followed seasonal patterns of availability in habitats ranging from extremely dry deserts to high-altitude alpine forests. In the southern Southwest, some groups experimented with husbandry of native annual plants that produced edible seeds. These experiments with sowing and soil enrichment initiated a process of coevolution between human and plant populations that was similar in many ways to the initial domestication of EAC crops. However, in most of the region it was not these native crops but maize, a domesticate introduced from Mesoamerica, that was central to initial farming efforts. Dietary reliance on maize varied considerably across time and space during the Early Agricultural period, but by 1,000 years ago it had become a staple throughout the Southwest. Meanwhile, in the East maize had made a tentative appearance over a millennium before, having arrived in Ohio, Tennessee, and southern Illinois by unknown routes and mechanisms. Nearly invisible archaeologically prior to AD 700, maize quickly ascended to a position of primary importance, largely eclipsing the crops of the EAC within centuries. The story of maize in the East is the topic of the next chapter.

4

The Rise of the Three Sisters: Maize in the Eastern Woodlands

In 1539, Hernando de Soto commenced his infamous tour of the North American Southeast. The Spanish entrada is well known for its violent encounters with Native peoples, which are described in the accounts of the Gentleman of Elvas and others (Hudson 1994). Not seeking colonization opportunities but rather wealth, *conquistadores* destroyed as they progressed. However, they were careful to preserve one of the most valuable of village assets: stored maize. The Spanish armies needed to be fed, and this was accomplished by outright theft if necessary. Everywhere there were fields of maize and full corn cribs, except where stories of devastation had preceded them, causing villages to be abandoned before the arrival of the Spanish. None of the Spanish accounts mention the plants we now recognize as members of the EAC, with the exception of the sunflower (Gremillion 2002b).

The dominance of maize also comes through consistently in accounts of French and English travelers and colonists from the sixteenth century onward. In the words of James Adair, trader and chronicler of the lifeways of the Creek and Cherokee people of the Southeast during the late eighteenth century, "Corn is their chief produce, and their main dependence" (Adair 1775:407). Observant chroniclers who were resident across multiple seasons might catch a glimpse of an ancient practice, like the broadcasting of seeds of *choupichoul* by the Natchez described by Le Page du Pratz (1972 [1758]; Smith 1992d) or the still-unidentified *melden* described by Thomas Harriot (1972 [1590]). However, it was maize that clearly dominated the agricultural systems of Native peoples from the northern limit of its cultivation to Florida and from the Atlantic Coast to the edge of the Great Plains. Despite considerable diversity in the investments made in particular crops and in food production generally, agriculture throughout the region was maize-based, with cucurbits (squashes, pumpkins, and hard-shelled gourds) and the common bean as secondary crops. Many other crops were cultivated for food and ceremonial use, but their contributions to human diets were minor in comparison to that of maize. Much had changed, then, in the food production systems of the East since the widespread adoption of EAC crops more than a millennium before the arrival of Europeans.

During that time, while maize-based farming was becoming widely established in the Southwest, maize was introduced to the East from sources in the Great Plains or beyond. It was adopted by EAC farmers, but did not develop quickly into a staple; instead, the archaeological signal of maize remains very weak until the interval AD 750–950, when rapid intensification seems to have occurred. This shift to maize-based diets is reflected in the macrobotanical record and human bone chemistry, and to a lesser extent in starch grains and phytoliths embedded in residues from cooking pots. The chronology of this shift from EAC crops to maize is remarkably consistent across the Eastern Woodlands. The coincidence of this transition with the emergence of Mississippian polities in the Midwest and Midsouth has generated much discussion of the possible causal relationships between agriculture, surplus production, monumental architecture, and social stratification. However, explanations for intensification of maize production and the long period of limited use that preceded it are numerous and varied, with no consensus in sight.

Culture Areas and Chronology

Maize became a staple across most of the Eastern Woodlands where growing conditions permitted. At its northern limits in what is now southern Canada, maize was often incorporated into subsistence as a minor dietary component. In some cases it was produced primarily for trade. To the West lay the Great Plains, the region that lies between eastern maize-growers and the southwestern sources of the crop. The role played by maize agriculture on the Plains was highly variable, ranging from the Plains Village tradition's maize-focused economy to the more wide-ranging foraging strategies of nonagricultural groups (Adair 2003; Adair and Drass 2011). Finally, in tropical and subtropical coastal zones skirting the Atlantic Ocean and Gulf of Mexico, reliance on marine and freshwater fish and shellfish persisted until the advent of forced labor regimes instituted by Spanish missions.

Eastern North America has a venerable tradition of archaeological taxonomy, resulting in a potentially bewildering array of drainage-specific units and sequences. For the most part I avoid these in the discussion to follow (but see Chapter 2 for the basic culture-historical framework for the East).

Timing, Routes, and Means of Introduction

The earliest widely accepted dates on eastern maize are from macroremains, mostly cob fragments, that have been directly radiocarbon dated (Table 4.1; Figure 4.1). Direct dating was made possible by the development of the accelerator mass spectrometry (AMS) method, which can yield accurate results with minute quantities of carbon (Smith 1992b). Using the AMS method to reassess some purported cases of early maize, Conard and colleagues (1983) showed that many of the samples from Illinois that were thought to be early based on stratigraphic association with dated charcoal turned out in fact to be much more recent. This reassessment emphasized the need for radiocarbon assays performed directly on the maize itself. As AMS dating became more widely available to archaeologists, it became the "gold standard" of evidence for first instances of domesticated plants. The push for direct dating of purportedly early maize resulted in the current consensus that it was widespread in the East as early as 300 BC.

A recent reanalysis of the dated material from the Holding site in Illinois (Simon 2017) demonstrates that direct dating is no panacea for the difficulties of constructing an accurate chronology for the initial adoption of maize. Of the 16 fragments of purported maize from Holding still available for

Table 4.1. The Two Earliest Dates Obtained Directly on Maize Macroremains from Archaeological Sites in Eastern North America.

Site	Location	Radiocarbon Years BP	Standard Error	Calibrated Age BC/AD (2-σ Range)[a]	Calibrated Median BC/AD	Source
Icehouse Bottom	eastern Tennessee	1775	100	AD 22–532	AD 252	Chapman and Crites 1987
Harness	south-central Ohio	1730	85	AD 86–534	AD 301	Fritz 1990
		1720	105	AD 76–548	AD 311	

[a]Calibrated using OxCal4.2, online version, https://c14.arch.ox.ac.uk/oxcal/OxCal.html; IntCal13 calibration curve. Sources: Bronk Ramsey (2009); Reimer et al. (2013).

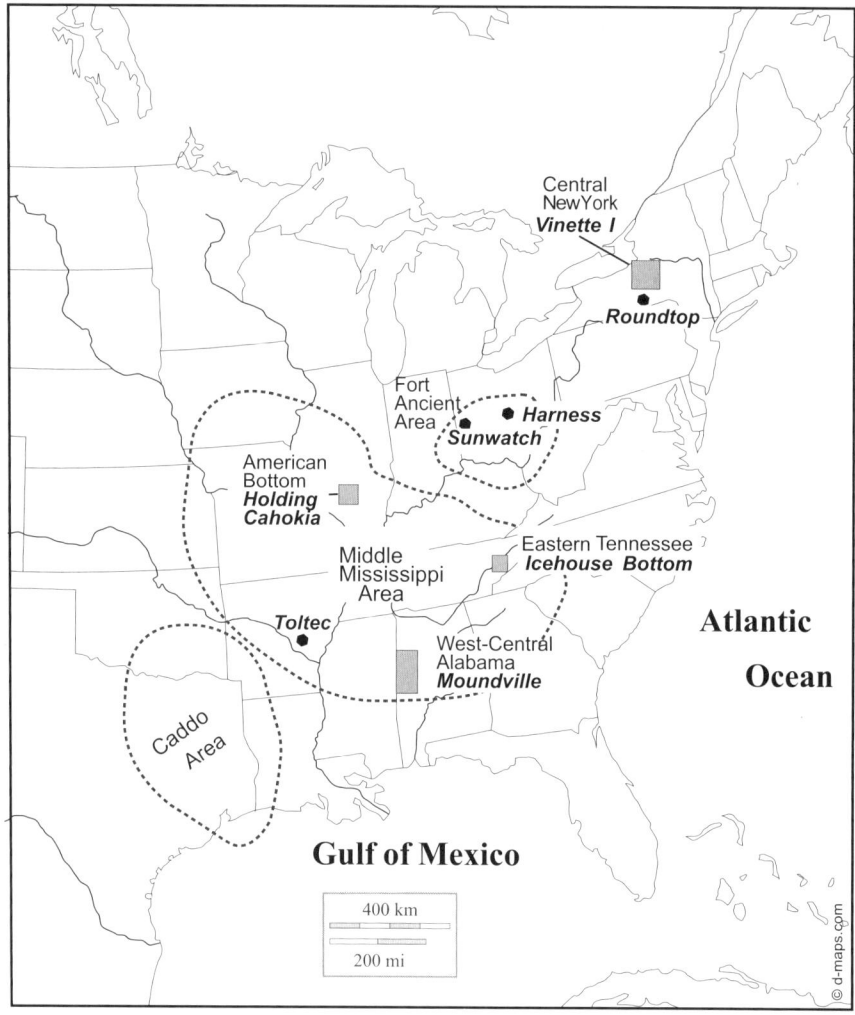

Figure 4.1. Map showing locations of sites and archaeological regions discussed in Chapter 4.

analysis, none that yielded Middle Woodland dates had been correctly identified. These small fragments were difficult to assess for distinctive cob and kernel morphology, but reanalysis of remaining specimens showed that they were not enriched in isotopic carbon (^{13}C), a diagnostic feature of maize. The materials from Holding that were correctly identified as maize proved to be much more recent than expected based on the original radiocarbon determinations. Thus there is reason to doubt entry of maize into southern

Illinois at this early date, requiring a reassessment of the apparently long period of low-level use alongside native seed crops there.

Recently researchers have begun to explore the potential of plant microremains for documenting the introduction of maize and other non-native crops to the northeastern United States and southern and central Canada. These microremains include starch grains and phytoliths (silica bodies that form in plant cells), which contain at best only minute quantities of carbon. Dating of occluded carbon within phytoliths is a promising technique, although it sometimes yields anomalous results (see discussion in Piperno 2016). However, phytoliths have the benefit of being resistant to decay, and both starch and phytoliths can be incorporated into the carbon-rich charred residues that are sometimes found inside cooking vessels.

These studies have yielded some surprisingly early evidence of maize preparation (and by inference, consumption) in the northern reaches of the Eastern Woodlands and the Plains. Based on starch and phytoliths of maize from cooking residues, maize became a widespread component of human diets on the eastern prairies of Canada as early as AD 700 (Boyd et al. 2006; Boyd et al. 2008). Maize use subsequently became widespread well beyond the northern limits of its cultivation, indicating that it must have been acquired through exchange (Boyd and Surette 2010). In central New York, charred food residues containing maize phytoliths from the Vinette I site date to cal 296 BC (Hart et al. 2007), which predates the earliest instance of maize macroremains at circa AD 500 (Hart 2014; Hart et al. 2007; Hart and Lovis 2012). This find reinforces the lack of any time-transgressive "wave of advance" of farmers or farming technology from southwest to northeast; the earliest dates on maize are also the farthest from the source areas. However maize was transmitted between groups, it dispersed too quickly to permit detailed reconstruction of its travels based on archaeological evidence.

For this reason, the route or routes of maize introduction remain obscure. In broad terms, maize was introduced from the Southwest and across the Great Plains into the Midwest and points east. The southwestern origins of eastern maize are supported by genetic studies of the variety Northern Flint, which was developed locally in the northeastern United States from a southwestern precursor (Doebley et al. 1986; Jaenicke-Després et al. 2003). En route to the Northeast, maize appeared on the central Plains no later than 500 BC (Adair 2003). Where maize apparently did *not* travel was along the shortest route from Mexico into the United States along the Texas Gulf Coast, which was occupied by hunter-gatherers throughout prehistory (Story 1985; White and Weinstein 2008).

The means and mechanisms of maize introduction are even less well understood than its chronology. Maize seems to have traveled into North America north of Mexico independently of other goods; there are no non-perishable Mesoamerican trade items associated with the relevant time period (Fritz 2011). The introduction of the plant into local economies is not associated with intrusive archaeological signatures of colonizing farmers nor with any accompanying specialized technology. Therefore, the most likely scenario is that maize traveled via down-the-line exchange from group to group, probably without metates or tortilla recipes.

This transmission pattern has implications for the costs of learning about maize, its preparation, processing, and growing requirements. If it was originally traded as a commodity, maize may have been uncoupled from the cultural knowledge associated with it at its source. In that case, experimentation would have been necessary to arrive at suitable planting methods (in hills rather than broadcast), diligent weeding, and protection from predators (Smith and Cowan 2003). Harvesting and processing would have presented challenges as well because the maize cob is different morphologically from the fruiting heads of the native plants that were cultivated for their seeds. The adherence of the kernels to the cob makes harvesting fairly simple, but kernels need to be removed from the cob unless they are consumed in an immature state as green corn. Flint and flour varieties require further processing to reduce kernels into smaller fragments for cooking. Treatment of maize with wood ash or some other alkaline substance, called nixtamalization, greatly improves its nutritional profile (Coe 1994). However, there is no evidence that this technique was introduced along with maize itself; instead, it was independently invented in the East after maize was introduced (Hart 2014).

The fact that maize seems to have traveled to the East unaccompanied by specialized technologies may help to explain why it was first adopted by farmers rather than non-farmers. Despite its unusual traits, maize had much in common with the starchy seeds, such as maygrass and goosefoot. Basic techniques for cultivating, harvesting, and processing annual seed crops were already part of the cultural repertoire of EAC farmers. This preexisting cultural knowledge would have reduced the time costs normally entailed by the process of learning effective ways to utilize this novel resource. Labor costs were limited by the small scale of maize production, which ensured that familiar strategies were not abandoned in favor of a crop whose yields were difficult to predict. Subsistence farmers are often risk averse, resisting development schemes that require wholesale adoption of agricultural innovations and abandonment of traditional methods despite promises of increased yields (Ghadim and Pannell 1999).

The Transition to Maize-Based Diets

For almost a thousand years following its introduction, maize contributed little to human diets in the East. This inference is based on the scarcity of macrobotanical evidence (fragments of cobs and kernels), but it receives even stronger support from the chemical analysis of human skeletal material. Unlike plant remains, which usually represent material discarded or lost during processing, human remains retain a chemical signature of the foods consumed. That signature is usually a general one indicative of broad categories of food that can be distinguished by the uptake of nutrients from different habitats or trophic levels (such as marine vs. terrestrial or animal vs. plant sources; see discussion in Chapter 3). Because maize is virtually the only plant with C_4 metabolism likely to have been present in the environment of the Eastern Woodlands, it is reliably indicated by elevated ^{13}C levels provided that the influence of marine foods can be ruled out (Ambrose 1987; Ambrose et al. 2003; Tykot 2006).

Carbon isotope studies have been conducted in several regions of the Eastern Woodlands—the Northeast and southeastern Canada, the Ohio Valley, the Southeast, much of the Mississippi Valley, and west-central and southern Illinois. These findings agree with the larger regional narrative drawn from archaeobotanical data in indicating the introduction of maize across a broad section of the midcontinent. This process was rapid enough to account for its presence in central New York State by 300 BC, but it is difficult to estimate the period of time involved because of uncertain chronological placement of maize macroremains from Plains sites (Adair 2003). First appearance of maize in archaeological contexts is typically followed by a long period of low-level use as a specialty crop (or perhaps in some areas its abandonment). A second episode of rapid change, which began nowhere earlier than AD 700, involved higher rates of maize consumption by more individuals. This second transition is inferred from the presence of individuals with ^{13}C values higher than about −20‰ (most values are negative with respect to the reference standard, the Pee Dee Belemnite). The average value for C_3 plants in the East, and the animals that feed on them, is −26.5‰, whereas the average for C4 plants is −12.5‰ (few animals with enriched ^{13}C were available for consumption; Emerson et al. 2005). However, ^{13}C values are not normally assessed in isolation, but rather relative to each other in a time series that should reflect any trends in average maize consumption.

Ongoing isotope studies have refined regional histories and identified some points of divergence from the larger narrative. A comparison of isotope trends from different regions, in conjunction with available archaeobotanical data, illustrates these variations on the theme of maize intensification.

The Caddo Area

In the Caddo area of eastern Texas, northern Louisiana, eastern Oklahoma, and Arkansas (Figure 4.1), maize was introduced by about AD 500 but did not become a significant dietary element until after AD 1200. Even then, it never achieved the degree of importance seen at Mississippian and Fort Ancient sites of the Midwest (Wilson and Perttula 2013). Within eastern Oklahoma, median ^{13}C values rise steadily starting at AD 1000, although with a great deal of variation between individuals of the same time period (Rogers 2011). Here on the western edge of the Eastern Woodlands, maize agriculture was gradually incorporated into mixed economies that continued to make substantial use of hunted and gathered foods as well as EAC crops.

Middle Ohio Valley

The transition to maize-centered diets in the middle Ohio Valley took place rapidly after AD 1000 (Greenlee 2006), when the archaeological culture known as Fort Ancient was in its early period of development (Cook and Price 2015). Greenlee's (2006) regional synthesis incorporates some collections that are not precisely dated, but nonetheless it is clear from this dataset that ^{13}C values higher than about −20 ‰ are found exclusively among individuals who died after AD 1000. Although some individuals obtained very little of their protein from maize after that time, their numbers are few. In a study focused on apatite values from human molars, Cook and Price (2015) were unable to detect any gradual increase in maize consumption at seven sites in southwestern Ohio and southeastern Indiana; in fact, they found some evidence for a decline in average ^{13}C values over time (for more on Fort Ancient, see "Variation within Communities," below).

The Northeast

In the Northeast, maize consumption is detectable in human remains by AD 500, and in subsequent centuries it became increasingly central in human diets. Its dietary contribution remained quite variable across the region and in some cases its cultivation was incorporated into high-mobility subsistence regimes (Chilton 2008; Hart and Lovis 2012). Skeletal collections from southern Ontario indicate a steady increase in the dietary contribution of maize following AD 500 (Crawford and Smith 2003; Crawford et al. 1997; Katzenberg et al. 1995). In coastal New England, carbon and nitrogen isotopes record small quantities of maize and a heavy marine diet between AD 900 and 1400 (Little and Schoeninger 1995).

The Southeast

In the Southeast and the lower Mississippi Valley, maize developed rather quickly into a staple crop. The archaeobotanical record for west-central Alabama, location of the Moundville polity, indicates that maize played a limited role until after AD 1000 (Scarry 1993). It is not known precisely when maize became available to the communities surrounding Moundville itself, but there is no evidence that it was introduced there as early as it was in the Northeast and Midwest. Perhaps the hunter-gatherers of the Southeast, who cultivated EAC crops on a small scale at most, faced a greater burden in terms of the costs of learning how to produce maize. In the coastal areas of Georgia and Florida, the role of maize was variable but in some areas at least a little maize was grown prior to the Spanish period, when missions pressured local Native groups to provide them with grain (Hutchinson et al. 1998). Most Florida populations show little evidence of maize consumption prior to the Historic period. In the lower Mississippi Valley, maize first appears between AD 1000 and 1200, but occurs in very low frequencies in the archaeobotanical record. Isotope data also support a limited dietary role for maize (Fritz and Kidder 1993). In the northern reaches of the Lower Valley at the Toltec Mounds site, intensive agriculture was practiced, but it was based primarily on native seed crops, with maize remaining a minor component.

The American Bottom and West-Central Illinois

In the Middle Mississippi area (including the American Bottom), the average level of maize consumption remained consistently high from AD 800 onward, with little change after AD 1000 (although in this case mean values obscure a great deal of variation between individuals, discussed further below). Pre-Mississippian skeletal samples from the American Bottom are few and generally poorly preserved (Hedman 2006), but at least some individuals in this part of the Midwest may have been eating maize-rich diets as early as AD 400 (Rose 2008). Although the skeletal material on which this interpretation is based has not been directly radiocarbon dated, archaeological context places it between AD 400 and 700. The influence of Cahokian agricultural traditions is also evident in northern Illinois, where maize was incorporated into an economy based largely on wild foods with small-scale cultivation of EAC crops. People of the Langford Tradition (AD 1100–1300) ate significant quantities of maize (about 70% of the whole diet), as did the Mississippian people of the American Bottom (Emerson et al. 2005).

Variation within Communities

As new research continues to clarify broad regional patterns, studies have turned increasingly to the examination of synchronic variation indicative of social status, gender, or immigration. Rose (2008) was able to show that there was considerable variability in the amount of maize consumed in west-central Illinois between individuals buried within the same mound. Fluoride dating demonstrates that this variation was not a function of time, but reflects differences between people likely to have been contemporaries. She documented a bimodal pattern of either high or low maize consumption, with few individuals falling into the intermediate range. Hedman's (2006) study of skeletal remains from the Cahokia area detected dietary differences between females and males at the East St. Louis Stone Quarry site. Her analysis makes use of the difference between bone collagen (which is organic and represents the protein component of the diet) and apatite (the mineral constituent of bone, which is more representative of whole diets). Based on the spacing between collagen and apatite values, males from the East St. Louis Stone Quarry site had relatively high ^{13}C levels that probably did not come from maize but from something else that was rich in protein, such as an animal that consumed C_4 plants. Another synchronic analysis from the middle Ohio Valley found a correlation between high ^{13}C values and archaeological indicators of Mississippian influence such as wall trench houses (Cook and Schurr 2009). Cook and Price (2015) found that, of the individuals at several Fort Ancient sites whose origins were nonlocal based on isotopic strontium levels, most were heavy maize consumers. However, both local and nonlocal samples showed much variability in levels of maize consumption that was not easily attributable to status or cultural connections to nearby Mississippian communities.

Maize and Human Health

The negative health effects of maize-based diets are well known and can be documented through study of skeletal pathologies in conjunction with dietary information (Larsen 1997). Before the advent of stable isotope studies, the frequency and severity of diet-related conditions such as iron-deficiency anemia and interruptions of growth during development were used as proxies for maize dependence. Analyses that combine carbon isotope data with information on skeletal pathologies show a clear correlation between maize consumption and developmental stress as well as indicators of infectious disease. In the lower Mississippi Valley, rates of dental caries are significantly higher in maize-dependent populations than those that preceded them, although in some cases starchy EAC crops account for the increased carbohydrate consumption that fuels bacterial infections (Rose et al. 1991). Iron-deficiency anemia has also been implicated as a consequence of increasing maize dependence, although bioarchaeologists recognize that the skeletal conditions caused by anemia (such as the porosity of bone known as cribra orbitalia or porotic hyperostosis) can also result from other causes, such as parasitic infections (Milner 1991).

The analysis of skeletal pathologies has led to important insights about status and its relation to diet. For example, Powell (1991) found little evidence of substantial differences between elite and non-elite burials with respect to skeletal indicators of stress during development (such as linear enamel hypoplasias), nutritional status, dental health, or infectious disease experience. She classified elite and non-elite individuals based on independent indicators such as location, form, and contents of burial pits. Powell's conclusions are based on a sample that did not include elite burials within mounds, which are expected to represent the uppermost strata of the Moundville community. However, her review of other comparative studies available at the time of publication (1991) shows that the Moundville pattern is unexceptional. Differences in maize consumption do not clearly differentiate between elite and non-elite individuals.

Summary

Rather than a gradual increase in economic importance, we see a long period of stasis anchored by two periods of rapid change, the first of which introduced maize into communities of forager-farmers across a broad geographical region. However, there is so little physical evidence of maize prior to about AD 750 that it is difficult to interpret these small scattered finds: Are they simply the tip of the iceberg of widespread, albeit small-scale, cultivation of the plant? Or do they instead record multiple short-lived introductions, importation as a commodity, or a pattern of use that was highly restricted? Whatever the nature of its presence in the East before AD 750, maize was not being consumed in quantities sufficient to create an isotopic signal. The second transition promoted maize from a specialty crop to a staple, a shift that had major impacts on human diet, health, and economic systems. The timing and rate of this change vary, but it started no earlier than AD 750. In most places where maize was adopted, it became a staple before the arrival of Europeans. It is not clear why this change happened when it did and not earlier, or indeed why it happened at all. But archaeologists have become particularly intrigued by the "lag" (a word choice that clearly alludes to an expectation of intensification as the eventual outcome), perhaps because maize-based economies are assumed to provide superior yields compared to the native seed crops and thus had utility that should have been readily perceived. In light of its later success as a crop, the relegation of maize to minor status for nearly a thousand years seems puzzling.

Explaining the Long Delay

Most of the explanations that have been proposed to account for the timing of adoption and intensification of maize in the East are functional, emphasizing a change in the profitability of maize production. These variously propose the development or introduction of improved, high-yielding maize varieties, or the development of more efficient processing methods. Other arguments cite a bias against preservation of maize at early sites that masks its true economic role, or argue that multiple introductions were needed to establish viable populations. These alternative explanations are discussed more fully in Chapter 7.

Improved Yields

Some researchers argue that maize required a long period of gradual adaptation to environmental conditions at northern latitudes, or that a high-yielding variety was introduced into the region late in prehistory. Maize originally evolved a flowering response that synchronized reproduction with the relatively short days of the rainy season in southern Mexico (Adams 2015). This photoperiodic trigger was maladaptive in northern latitudes because it delayed flowering until the end of the growing season, leading to a reduction in daylength sensitivity in temperate varieties (Coles et al. 2010).

Technological Innovation

The profitability of maize farming can also be increased by technological innovations that reduce the costs of processing or increase nutrient availability. Perhaps the best known of these techniques is nixtamalization, in which wood ash or some other alkaline substance is cooked with maize to produce hominy (Coe 1994; Hart and Lovis 2012). This process makes the essential amino acids tryptophan and lysine more bioavailable (see Chapter 7).

Differential Preservation Effects

Functional explanations assume that the "lag" in archaeobotanical evidence accurately documents a long period of widespread but limited production. However, as John Hart (2014) has pointed out, the small quantities of maize macroremains that predate AD 750 probably greatly underestimate its actual occurrence in the archaeological record. That record may also be biased by consumption of maize in its immature state ("green corn") or as a source of stalk sugar, uses that were not conducive to preservation through carbonization (Hart 2014). If these biases affected earlier periods of maize use more than later ones, then our understanding of maize history may be significantly altered as more researchers seek alternative lines of evidence.

Multiple Introductions

If preservation biases are not to blame for the low archaeological profile of maize until AD 1000, then perhaps the spottiness of the record reflects multiple introductions. Because maize is entirely dependent on humans for its propagation, its spread hinged on opportunities for social interaction, at the very least involving a transfer of goods in the form of seed corn. Small populations of maize are difficult to maintain because of the relatively short dispersal distance of its pollen and consequent inbreeding depression (Hart 1999, 2008). Many experiments with maize cultivation probably failed to establish persistent populations, contributing to the discontinuous archaeobotanical record and accounting for its scarcity.

Maize-Based Agricultural Systems

The economic role of maize production was highly variable across the Eastern Woodlands. Although the Late Prehistoric period (roughly from AD 1000 to European contact) is usually characterized in terms of maize dependency, the degree of dependency, the importance of other crops both imported and indigenous, and even the mode of acquisition (cultivation or exchange) were not homogeneous across populations. Early observations by European travelers along the Atlantic coast helped to form the classic image of the "Three Sisters" system (maize, beans, and squash grown together as a distinctive and ecologically robust resource patch). As valuable as these accounts are, they are incomplete and imply a uniformity that archaeological and ethnographic research has failed to support.

Although the complementarity of the Three Sisters is reasonably well understood in ecological terms (Mt. Pleasant 2006), other aspects of maize-based production systems in prehistory remain somewhat elusive. Considerable disagreement exists, for example, regarding the prevalence or even presence of true swidden cultivation, in which small plots are deforested (with the assistance of fire and cutting tools), used for a brief period, and then abandoned in favor of a new site (Doolittle 2000). Many historic accounts indicate greater permanence of cultivated gardens or fields, but villages were still periodically relocated in the face of dwindling supplies of game, fuelwood, and fertile soils within local catchments. Archaeological documentation of production systems is hampered by the difficulty of locating ancient fields and the general scarcity of persistent landscape modifications such as ridges and hills. Water management features such as the stone mulch, check dams, channels, and canals so common in the Southwest are less useful in the temperate climate of the East. However, several case studies illustrate the impact of settled farming populations on local vegetation as documented by assemblages of plant remains. There is also good evidence for the non-maize crops introduced from more southerly sources—common bean, a squash (*Cucurbita mixta*), domesticated forms of chenopod and amaranth, tobacco—and of the variable fate of EAC crops in the maize era (Fritz 2011; Lopinot 1992).

The Three Sisters

In the Northeast, maize, beans, and squash were grown together wherever they were found (Hart 2008). The earliest archaeological evidence of this phenomenon comes from the Roundtop site in New York, where remains of the three crops together produced a pooled mean uncalibrated date of 667 ± 30 BP (cal 2-σ range AD 1276–1391; Hart 2008). The common bean (*Phaseolus vulgaris* L.) was domesticated at least twice, once in Peru on the eastern slopes of the Andes, and once in west-central Mexico north of the Balsas River valley where teosinte first became maize. In Mexico, there is evidence of common bean from the caves of the Tehuacán and Oaxaca Valleys by about 350 BC. In North America, beans appeared east of the Mississippi no earlier than AD 1300 based on radiocarbon assays of carbonized fragments of cotyledons (embryonic leaves; Brown et al. 2014; Fritz 2011). The spread of the common bean to northern latitudes may have been slowed by the plant's sensitivity to long days, which tended to delay flowering and truncate the growing season (Coles et al. 2010; Hart 2014). The third member of the Three Sisters complex, *Cucurbita pepo*, has a lengthy history in the East. *C. pepo* ssp. *ovifera* is native to the eastern United States and was domesticated there beginning as early as 3000 BC (see Chapter 2). This native eastern squash is believed to have been adapted to dispersal by megafauna, a role that humans took over in the wake of the late Pleistocene extinction event that removed most of the largest herbivores (Kistler et al. 2015). The Mexican lineage *Cucurbita pepo* var. *pepo*, which was domesticated in central or southern Mexico some 10,000 years ago, is best known as the jack-o-lantern pumpkin (Fritz 2011). Although cultivars of var. *pepo* were transmitted to the Southwest prehistorically, they left no definitive traces in the East prior to European contact (although the condition of remains makes it difficult or impossible to distinguish Mexican from native eastern North American squashes).

The Three Sisters crop complex formed the core of a "sophisticated and rational" (Mt. Pleasant 2006:530) system of soil and crop management. In this system, maize kernels were planted in small hills, several at a time. Beans were allowed to vine up the cornstalks, and squash was grown in swales between the hills. The leaves of squash plants create a living mulch that helps to retain soil moisture during the hot, dry summer months. Intercropping in this way reduces pest damage and exploits the complementarity of legumes (which fix nitrogen in the soil) and maize (which is a heavy consumer of nitrogen). The function of mounding soil is contested, but it seems likely that mounds and ridges helped to protect plants from frost damage and created warmer temperatures for spring planting (Mt. Pleasant 2006). This practice is effective because cold air tends to gather in depressions, causing an inversion of the more common pattern of decreasing temperature with increasing altitude. Mounds also helped to improve soil characteristics by facilitating drainage and controlling the spacing between plants, which is essential to maximize yields because maize does a poor job of generating "tillers" to fill in unoccupied space. Mounds were enriched by the addition of harvest by-products, concentrating organic matter and supplying nutrients, and plants growing on mounds were protected from soil compaction caused by trampling (Mt. Pleasant 2006).

Other Crops

The Three Sisters system has been documented archaeologically in the Northeast and the Great Lakes region (Mt. Pleasant 2006). Elsewhere in the East, squash and beans were less central and production focused more on maize, for example in the middle Ohio Valley and the Southeast. In the American Bottom, common beans seem never to have become a very important crop, although one native legume, the groundnut, is often found in archaeological contexts (Lopinot 1992). In contrast, the common bean was important in the upper Mississippi Valley and in the middle Ohio Valley during the Late Prehistoric period. So although maize, beans, and squash were grown throughout the East, the members of the trio played roles that varied with local ecologies and histories.

The type and number of other crops that made a lesser contribution to human diets also varied greatly. The Fort Ancient area is known for its heavy reliance on maize and its near-abandonment of EAC crops after AD 1000 (Wagner 1987), in contrast to the American Bottom, where native seed crops were not replaced until late in prehistory (Lopinot 1992). In fact, indigenous cultigens such as goosefoot and maygrass may have been shifted from a central economic role to a more ceremonial one. The importance of EAC crops in communal meals is illustrated by the inclusion of these plants within a submound borrow pit at the Cahokia site. This pit, which was filled in discrete episodes during a period of several years at most, contained unexpected fauna (e.g., swans) and unusually abundant bones of large river fish. Among the 50 or so plant taxa found in this uncarbonized assemblage were numerous seeds of squash and native seed crops. Other archaeological evidence supports their deposition in the contexts of a communal meal associated with mound construction (Pauketat et al. 2002).

Aside from the Three Sisters, few non-native crops were adopted in the East between the arrival of maize and European contact. One of these introductions is a domesticated form of amaranth native to Mexico, *Amaranthus hypochondriacus*, which has been identified in dry deposits from rockshelters in northwestern Arkansas dating to AD 1000 (Fritz 1984a, 2011). This pseudocereal has small grains and a thin seed coat, which is also a defining characteristic of certain cultivars of the closely related genus *Chenopodium* (see Chapter 2). Despite its obvious role as a stored crop in some contexts, grain amaranth does not seem to have been geographically widespread in the East.

The other introduced plant is tobacco, which has a complex history in North America. The earliest archaeological examples consist of seeds dated to AD 100 and nicotine-containing residue from an Adena pipe that is probably several hundred years older (Dunavan and Jones 2011; Fritz 2011; Rafferty 2006). Individual seeds of tobacco, usually less than 1 mm in diameter, were unlikely to have been found prior to the era of flotation recovery using fine-mesh screens. However, since that time there have been many such finds from Woodland and Late Prehistoric period contexts (Rafferty 2006). The tobacco first encountered by European explorers on the Atlantic Coast was long thought to have been *Nicotiana rustica*, a species with high nicotine content that ultimately originated in South America. Once believed to account for all archaeological instances of tobacco in the East, *N. rustica* may have been accompanied or even preceded by native tobaccos of the West such as *N. quadrivalvis* (Wagner 2000). However, it is difficult to differentiate the species morphologically based on seed coat patterning, particularly in charred and damaged specimens (Dunavan and Jones 2011; Fritz 2011).

Maize-Based Farming Systems and Ecological Impacts

Knowledge of how eastern North American maize farmers cleared and managed their fields has been hampered by the lack of direct evidence of ancient fields and gardens (Gallagher and Arzigian 1994). Paleosols that indicate ancient fields, even if preserved, may lie under deep alluvial deposits. Some field remnants exist in the form of visible ridges and other topographic features, most of which are located in the Upper Midwest, but these are relatively rare. The archaeological evidence we do have consists mainly of plant remains (including phytoliths, pollen, and macroremains) that are proxies for vegetation changes such as removal of the forest canopy. Inferences can be made about agricultural technology based upon artifact forms and frequencies. Zooarchaeological assemblages sometimes document increases in commensal animal species associated with forest clearing. The fact of food production itself tells us little, and even quantities of material residue from the harvesting and processing of crops can only suggest the degree of labor investment. The ethnohistoric and ethnographic records provide intriguing hints of actual practice, but these tend to be brief and scattered in the literature. Relatively few of these accounts date to the period of initial European contact and thus cannot be considered indicative of precontact patterns.

Discussions of maize-centered agriculture in the East usually incorporate comparisons with ethnographically documented systems. However, inferences based upon analogy in this context are underdetermined by material evidence, which is often consistent with a range of possibilities much larger than that defined by existing documentation. Despite this drawback, well-studied agricultural systems provide a useful framework given the paucity of direct evidence for prehistoric strategies of clearing, planting, maintaining, and harvesting gardens or fields.

MAIZE YIELDS. The question of maximum yields is an important one because of its implications for the accumulation of surplus, which can provide a buffer against food shortage or be mobilized for social display and communal activities. The yields of maize itself can vary considerably depending on soil moisture, fertility, and other environmental factors. There may also have been changes in average yields contingent on the development of landraces adapted to local conditions. The superiority of maize yields over that of EAC crops is often cited as a major factor in maize adoption and eventual intensification, although it seems rather unlikely that the earliest varieties adopted were vastly superior to existing crops. Smith (1987a) estimated productivity in free-living stands of *Chenopodium berlandieri* and found that total yields overlapped with those estimated for maize, although maximum yields of maize are somewhat higher when yields of goosefoot are adjusted to reflect the indigestibility of the seed coat. The other EAC crops assessed in Smith's study had yields lower than or comparable to goosefoot. By Smith's estimate, maize would have had to exceed yields of 1000 kg/ha (based on statistics from upper Missouri River Native communities of the 1860s and 1870s) in order to out-produce goosefoot. Schroeder (1999, 2001) argues that reliance on nineteenth-century data artificially inflates maize yields relative to what might have been produced in precontact times because of the adoption of European-style farming using plows, the practice of polycropping, post-harvest losses due to spoiling, and insect predation. Her revised estimate of maize productivity is on the order of 600 kg/ha on average, comparable to values estimated for goosefoot (Smith 1987a). A more realistic model that incorporates soil characteristics and yield estimates for specific cultivars, although couched in the form of a critique of Schroeder's methods, came up with a nearly identical estimate for average yields (approximately 500–750 kg/ha) while emphasizing that much higher values up to 1880 kg/ha might be expected in exceptionally good years (Baden and Beekman 2001).

Whether maize offered superior yields to EAC crops, and if so under what conditions, remains unclear. All estimates of this kind are hampered by uncertainty. However, the debate indicates that the economic advantages offered by maize-based farming over native seed crop cultivation cannot be taken for granted. Perhaps maize increased the upper limit of productivity for plant foods, but yield maximization would have come at the cost of relinquishing some of the crop diversity that helps to limit variance in production. There may have been other constraints on maize intensification, such as the difficulty of maintaining small populations in isolation.

CROPPING SYSTEMS AND LAND USE. The archaeobotanical record yields information about which crops were grown in fields and gardens along with maize. Site locations can be used to infer decisions about field placement and soil characteristics. However, many aspects of agricultural systems remain obscured by the limitations of the archaeological record, including the size and permanence of fields, methods of clearing forested land, and even whether maize and other crops were grown primarily as monocultures or in polycropped, structurally complex gardens. These aspects of agricultural systems have been derived from historical documents, particularly the accounts of European explorers and travelers who observed landscapes and cultural practices that were largely free of European influence (with the exception, of course, of European pathogens, which spread through Native communities before direct contact had become routine). Though imperfect and partial, historical accounts augment archaeological data by providing key information on how maize farming was practiced on the cusp of the Historic period.

It is clear that Late Prehistoric peoples of the East maintained a diverse system of food production in most areas despite the dominance of maize (see discussion above under "Other Crops"). With the exception of the Three Sisters, relatively little is known about how different crops were combined in mixed plots. Monocropping was certainly practiced; early explorers, most notably the Spanish who traveled the interior Southeast in the 1540s, wrote of extensive maize fields, in some cases quite far from settlements (Elvas 1993). Some of these were said to be communal fields cultivated to sponsor feasts or for contributions to the chief's stores (Scarry and Scarry 2005). Smaller gardens nearer to houses were maintained by households for their own use, at least in the Southeast (Adair 1775; Hatley 1989).

SHIFTING CULTIVATION VERSUS PERMANENT FIELDS. There has been much discussion in the literature of whether or not Native eastern North Americans practiced a form of swidden cultivation. The geographer William Doolittle argues against this interpretation, claiming that large fields were cultivated on a permanent basis (Doolittle 2004). Although fields were cleared using a combination of cutting and burning, he says, there is little evidence of fallowing (allowing fields to "rest" after harvest) or field rotation. He cites as evidence for permanent cultivation the fact that early explorers noted many "abandoned fields" that were not being replanted, although he also acknowledges that depopulation was probably partly responsible for field abandonment. He cites as further evidence the fact that removal of trees was difficult and therefore too costly to be undertaken with the frequency required for swidden cultivation. Even stumps were removed from fields, which is usually not done for fields intended for only a few years of use. In the temperate climate, in contrast to the tropics, reforestation was a slow process and the time required for recovery too long to warrant frequent field relocation.

The contrast between active, fertile maize fields and depleted, useless fallow land should not be overdrawn, however. "Old fields" remained productive even after cropping ceased. They were invaded initially by annual weeds and later by perennial shrubs and disturbance-loving trees, many of which produce edible seeds or fruits. Although they appeared unkempt to European eyes, old fields with fruit trees such as plums and persimmons were reminiscent of orchards (Fritz 2000; Hammett 2000). Fallow fields were also attractive to many game animals and therefore had value as "garden hunting" venues (Hammett 2000).

Although fields may have remained under cultivation for longer periods than is typical in tropical systems, they inevitably yielded diminishing returns over time. Foster (2003) examined the economic implications of declining yields, which according to historical records often motivated a shift to a new location. He created a quantitative model of agricultural production by Muskogee people of the village of Cussetuh in Georgia in the late eighteenth and early nineteenth centuries. He estimated yields for specific fields used during the occupation of the village and assumed that fields would have been abandoned when they ceased to produce the minimum required to feed the population a 55% maize diet. The model predicted that frequent field relocation would have been necessary in order to meet these needs, resulting in fluctuating yields. Historic records do in fact document field relocation in response to inadequate yields. Foster's assessment is that it would have been possible to rotate field locations within the larger area claimed by the town until, after about 30 years, fallow periods would have become so short (on the order of a single year) that the land could no longer support the population. This pattern of frequent field relocation within a larger catchment resembles tropical swidden systems in relying on switching field locations rather than crop rotation or soil amendments as a means of maintaining productivity.

TOPOGRAPHIC MODIFICATIONS. There are but a few examples in the East of persistent topographic or hydrologic features designed to buffer crops from extreme environmental conditions. Unlike the Southwest, the East normally receives adequate rainfall for successful cultivation of maize, although periods of drought were not uncommon (Anderson et al. 1995; Benson 2009; Munoz et al. 2015). There is no indication of intensive cultivation of wetlands that might have involved the construction of canals and drainage ditches. Presumably there were other options available to farming communities; uplands could be and were farmed when the preferred alluvial soils were unavailable. The Mill Creek chert hoe blades that were widely traded during Mississippian times in the American Bottom made this task feasible when floodplains were too wet to cultivate (Hammerstedt and Hughes 2015). These hoes were capable of breaking up root-bound prairie soils.

It was common throughout the East to elevate maize plants above the surrounding land surface, whether by planting the kernels in small hills or mounding earth up around the base of young maize plants. More extensive topographic modifications were sometimes undertaken where conditions were marginal for maize cultivation, and remnants of these are present today in the upper Mississippi Valley. The ridged fields of Wisconsin and Michigan take many forms, but usually consist of sets of parallel linear ridges constructed (presumably) to modify microclimates around maize plants. Elevating the plants above the level of the field would have created a temperature inversion, allowing colder air to sink into the furrows, much in the same way corn hills did for Three Sisters agriculture in the Northeast. Ridged fields also improved drainage in flood-prone fields, an important risk reduction strategy where the short growing season did not permit a second planting. Ridged fields therefore reduced the likelihood that a crop would be lost to late frost or spring flooding (Gallagher and Arzigian 1994).

ENVIRONMENTAL IMPACTS. Historical accounts from the interior Southeast reveal an anthropogenic landscape. Besides the extensive forests that so impressed Europeans, the vegetational mosaic included old fields at different stages of ecological succession (Hammett 2000). Although they may have been fallowed with the intention of future replanting, these patches also provided resources in the form of annual fruit- and seed-producing plants. Active regrowth in the wake of burning or other methods of vegetation removal attracted browsing animals, such as whitetail deer.

Native peoples maintained this "shifting mosaic" through activities such as prescribed burning, agricultural clearing, and managing fruit and nut trees by removing competitors and maximizing canopy area (and thus productivity; Hammett 2000). There is disagreement regarding the frequency and intensity of controlled burns in the East, but there are many accounts that mention the use of fire to clear forests and fallow fields for cultivation. In New England, prescribed burns were done once or twice per year to stimulate sprouting, a practice that also kept forests clear of accumulated woody material that could fuel damaging forest fires (Day 1953). Oaks and other fire-tolerant trees were not killed by low-intensity fires; in fact, they often enjoyed increased production of fruits (including acorns and other tree nuts used for food) with the removal of competing vegetation (Hammett 2000).

Even where high-intensity fires killed trees, forests would eventually recover, but more lasting changes accompanied the establishment of populous villages. For example, in the American Bottom, charcoal assemblages document decreasing quantities of fuel from bottomland trees and a corresponding increase in upland woods. This pattern is interpreted to reflect selective deforestation of riverine habitats for farming, forcing residents to turn to upland sources of firewood (Rindos and Johannessen 1991; Woods 2004). Deforestation ultimately increased rates of erosion and flooding during the growing season, although a period of unusually wet climate may have also contributed (Benson 2009). Environmental degradation was severe by the time the population at Cahokia reached its peak at AD 1050 and is thought to have stimulated expansion into the less fertile prairie soils found above the bluffs that fringe the Mississippi floodplain (Hammerstedt and Hughes 2015; Woods 2004).

A similar pattern of impact on bottomland forests is documented by pollen and wood charcoal data from the lower Little Tennessee River valley in eastern Tennessee. Charcoal of bottomland taxa declines after about 2000 BC, a pattern that probably reflects the decrease in land area with erosion but may also be related to clearing and cultivation. Upland taxa and woody plants adapted to disturbance make up the bulk of wood charcoal assemblages thereafter, with the latter comprising about half the total during Mississippian times (Chapman et al. 1982).

Maize and the Mississippian

In the Midwest and the Southeast, Mississippian polities were forming around the same time maize was making the transition from minor to staple crop. Mississippian cultures show evidence of emerging or persistent social stratification, some degree of political centralization, and a geographically widespread shared system of meaning displayed in iconography and ritual performances (Brown 1994). Mississippian peoples were maize-centered in their agricultural practice, but still maintained a wide range of foraged resources that contributed significantly to their diets (Smith 2009). Other Late Prehistoric maize-centered economies existed on the geographical fringes of the Mississippian world, for example, Oneota in the upper Mississippi Valley (Hart 1990) and Fort Ancient in the middle and upper Ohio Valley (Wagner 1987).

It was once believed that maize production fueled the development of social complexity and hierarchical organization among Mississippian peoples. In fact, archaeologists of the early twentieth century assumed that *only* maize agriculture could have provided adequate surplus to fuel differential accumulation of wealth and power by individuals and kin groups. The discovery of substantial earthen monuments dating back to the Middle and Late Archaic periods (Saunders et al. 1997) has debunked the notion that maize farming (or indeed, farming of any kind) was a prerequisite for moundbuilding. However, there is a clear association in time between the growing importance of maize in subsistence economies and the development of densely populated communities with evidence of defined public space and an increasingly hierarchical social order. Many questions remain surrounding the causal relationships between maize intensification, accumulation of surplus, and social inequality.

A common theme in discussions of maize intensification and the Mississippian is the mobilization of surplus for status competition, rewarding political allies, and creating social obligations. Evidence for these activities has been found in the form of residues of communal meals, some of which included rarely used foods (Blitz 1993; Jackson and Scott 2003; Pauketat et al. 2002). Feasting is indicated by large serving vessels such as those recovered from mound contexts at Lubbub Creek in west-central Alabama along with a faunal assemblage that contained a number of bird species rare in domestic contexts (Blitz 1993). In the vicinity of nearby Moundville, ceramic evidence of feasting increases during the Mississippian period along with the frequency of extremely large storage jars (Barrier 2011; Scarry and Scarry 2005). However, whether Mississippian elites controlled both access to stored food and its distribution through feasting and gift-giving has been contested (Barrier 2011). Understanding of the relationship between maize production and social complexity is currently moving beyond the chiefly redistribution scenarios popular in the late twentieth century.

New methods in geochemistry are proving to be a powerful tool for exploring the role of Mississippian immigrants in the dispersal of maize. Analysis of human skeletal remains from Fort Ancient sites reveals that individuals whose isotopic strontium profiles identify them as originating in the

Mississippian culture area (Figure 4.1) tended to consume more maize than did locals (Cook and Schurr 2009). High levels of maize consumption as measured by ^{13}C are also associated with Mississippian architecture and artifacts. Several human burials at the Sunwatch site, a large Fort Ancient village in Ohio, that contain unusually high levels of ^{13}C are associated with a Mississippian-style wall trench house. Further support for in-migration comes from isotopic strontium in dental enamel from burials at several Fort Ancient sites, which shows that nonlocal individuals from Mississippian territory were regular consumers of maize (Cook and Price 2015). Clearly the relationship between maize and the Mississippian is much more complex than previously supposed.

Summary

Maize-centered agriculture became normative across much of eastern North America during the last millennium. Starting around AD 750, human groups in the midcontinent began a fairly rapid transition from a mixed economy that included small-scale, seasonal use of cultivated seed crops to one that was still mixed but heavily agricultural and dependent upon maize production for both immediate and future consumption. The success and spread of these maize-focused economies was made possible in part by the accumulation of ecological knowledge across millennia of residence in temperate forest habitats. That knowledge was the basis for ecological management practices that were widespread well before domesticates entered the picture and persisted until recent times. Such practices represent forms of intensification that enhanced the yields and predictability of key resources, even where agriculture sensu stricto failed to take hold. These food producers without agriculture are the subject of the next chapter.

5
Food Production without Farming

Thus far, this book has been occupied with the lifeways of people who can be labeled without controversy as farmers. They meet all the criteria for the practice of agriculture described in the introduction to this book: economic intensification; use of domesticates; reliance on domesticates to provide a large proportion of food energy; labor investment that diverts energy from alternative activities; and persistent anthropogenic habitats. These criteria are clearly met by the maize-growing societies of late prehistory, in both the Southwest and the East. Furthermore, these groups relied on seed crops planted in fields, tended during the growing season, and harvested for both immediate consumption and storage. The position of EAC cultivators on the food production continuum is harder to determine because of uncertainty about labor investments, dietary importance of food crops, and their cultural significance. However, their use of domesticates and creation of anthropogenic habitats certainly qualify them as low-level food producers with domesticates (sensu Smith 2005b; see Chapter 1 for an inclusive definition of food production).

This chapter, however, is about groups that have long been classified as hunter-gatherers although they are more accurately described as low-level food producers. The astute reader may question its inclusion, with some justification. Why not simply expand the depth or breadth of coverage of intensive agricultural systems? However, there are sound reasons for going beyond conventional boundaries in this way (see Chapter 1). Classification of subsistence economies is a messy business and can be a hindrance to understanding. Even societies that practice food production without domesticates, or that engage in agricultural activities on a sporadic or opportunistic basis, have an impact on the habitats on which they depend. In some cases, they manage enhanced resource patches and practice intensification in ways that make them similar to farmers in residential stability, social complexity, and reliance on food storage. They have much to teach us about the diversity of subsistence adaptations in North America.

A second justification for this chapter is that this diversity of subsistence adaptations demands explanation. Why did some North American populations remain closer to the "food procurement" end of the continuum (Smith 2001)? In some cases, environmental constraints preclude production of most food crops. However, in others agricultural dependence developed despite considerable environmental challenges; in some locations it did not, even where conditions were relatively benign for growing crops. The "agricultural" end of the continuum is not an inevitability, nor is it the only alternative to foraging. Incorporating these diverse outcomes into the discussion of food production provides the comparative perspective required to evaluate causal pathways involved in subsistence change.

Reconsideration of the forager-farmer dichotomy has opened up new avenues of research that support an inclusive definition of food production. Two case studies will illustrate how food production can play an important role in economies that have little or no involvement with the seed crops and cultivated fields that have come to exemplify agricultural subsistence. Societies of the Northwest Coast of North America were for many years regarded as anomalous because their foraging economies seemed incompatible with social complexity. Cultural elements such as social ranking and wealth inequality, support of specialists, and dense populations were believed to depend upon agriculture as a reliable source of surplus. More recent research on the use of plant resources by the societies of the Northwest Coast and adjacent Inland Plateau (which together make up the Pacific Northwest), however, clearly demonstrates that they put considerable effort into enhancing the growth of perennial plants. The second case study targets the Great Basin, whose high-elevation deserts are marginal at best for agriculture. In prehistory the region supported mobile foragers with a broad-based subsistence adaptation based on wild animals and plants. Part of that adaptation included the sowing and harvesting of non-domesticated grasses, practices not usually associated with high mobility. Peoples of the southern Great Basin also found ways to incorporate maize into their diets without abandoning entirely their mobility and traditional foods. Much of the evidence from both regions is ethnographic, but they also have a long history of archaeological research focused on subsistence, particularly in the Great Basin.

Why have anthropologists been so slow to recognize the importance of food production to these populations? The persistence of the forager-farmer dichotomy has contributed to the tendency to categorize low-level food producers as hunter-gatherers by default (as not-farmers) despite being recognizably different from agriculturalists (Smith 2001, 2005b). Another contributing factor is the model of cultural evolution as a predictable sequence of stages in which markers of social complexity are inevitably associated with a secure agricultural subsistence base. This assumption, although clearly out of favor with theoretically minded archaeologists, seems to persist in implicit form in some published work. Ethnohistoric accounts are problematic because European observers saw agriculture in terms they were familiar with, as the cultivation of permanent fields of cereal crops coupled with animal husbandry. Maize farmers were at least somewhat compatible with the European experience, but the visitors also encountered groups that grew domesticated plants in an opportunistic way, or cultivated non-domesticated plants, or managed using methods (such as controlled burning) that Europeans regarded as destructive.

The Pacific Northwest: Clam Gardens and Camas Meadows

The Pacific Northwest, sometimes called "Cascadia" (Ames 2005), includes both the Northwest Coast and adjacent portions of the Columbia Plateau. The Northwest Coast proper has a special place in the history of anthropology because its peoples played a formative role in the thought of early twentieth century anthropologists, including Franz Boas. Relatively little of the large body of ethnographic research devoted to the region in the first half of the twentieth century, however, related to subsistence patterns; the cultural anthropology of the era was more interested in belief and ideology, ceremony and ritual, oral traditions, and social organization.

Boas rejected evolutionary models of cultural progress in favor of a strongly empirical approach in which detailed local histories offered the best way to understand culture change (Harris 1968). However, cultural anthropology took an evolutionary turn with the revival of cultural evolutionism by scholars such as Elman Service, Marshall Sahlins, and Leslie White (Sahlins and Service 1960; Service 1975; White 1959). While acknowledging "special evolution" as a historical process, these evolutionists expected "general evolution" to conform to a predictable sequence in which agriculture was the foundation of complex societies (chiefdoms and states). In this framework, Northwest Coast cultures were anomalous because they had most of the traits of the chiefdom level of organization but lacked agriculture. This apparent paradox was explained as an unusual case of superabundant natural resources, particularly salmon and other anadromous fish. These, it was believed, made agriculture unnecessary as a source of surplus that could be mobilized for status competition and alliance formation. Plant foods were seldom factored in because they had relatively low visibility compared to salmon and other aquatic resources. Practices that increased the yields or reliability of wild plants were either unrecognized or regarded as being of only minor importance (Ames 2005).

Recent research (see especially chapters in Deur and Turner 2005) makes a strong case that peoples of the Pacific Northwest are, and were historically and prehistorically, active food producers. Many of the target species are plants, which have traditionally been deemphasized by researchers in favor of the superabundant animal foods (such as salmon) regarded as the primary basis of subsistence. In fact, intensification of plant resources underwrites the "natural" abundance so often attributed to the Pacific Northwest region (Weiser and Lepofsky 2009).

The Pacific Northwest as Human Habitat

The Pacific Northwest is environmentally diverse and includes parts of the Western Forested Mountains, Marine West Coast Forests, and North American Deserts ecoregions (see Chapter 1). Topography and vegetation vary considerably across this region, which includes coastal and montane forests as well as dry coniferous forests, shrub-steppe, and grasslands in the interior (Butler and Campbell 2004; see also Figure 1.1). The Cascade range runs north-south (Figure 5.1), dividing the region into a well-watered eastern area with milder temperatures and a western area with lower precipitation that lies in the mountains' rainshadow (Deur and Turner 2005). The western area supports dense forests, while east of the mountains coniferous forests and prairies dominate. The mountains that separate the eastern and western areas feature a series of vertical habitat zones that culminate in alpine vegetation above the treeline.

The diverse set of habitats and distinct microclimates create a highly patchy landscape in which a wide range of resources can be reached within a small area (Deur and Turner 2005). Terrestrial productivity is high, but most

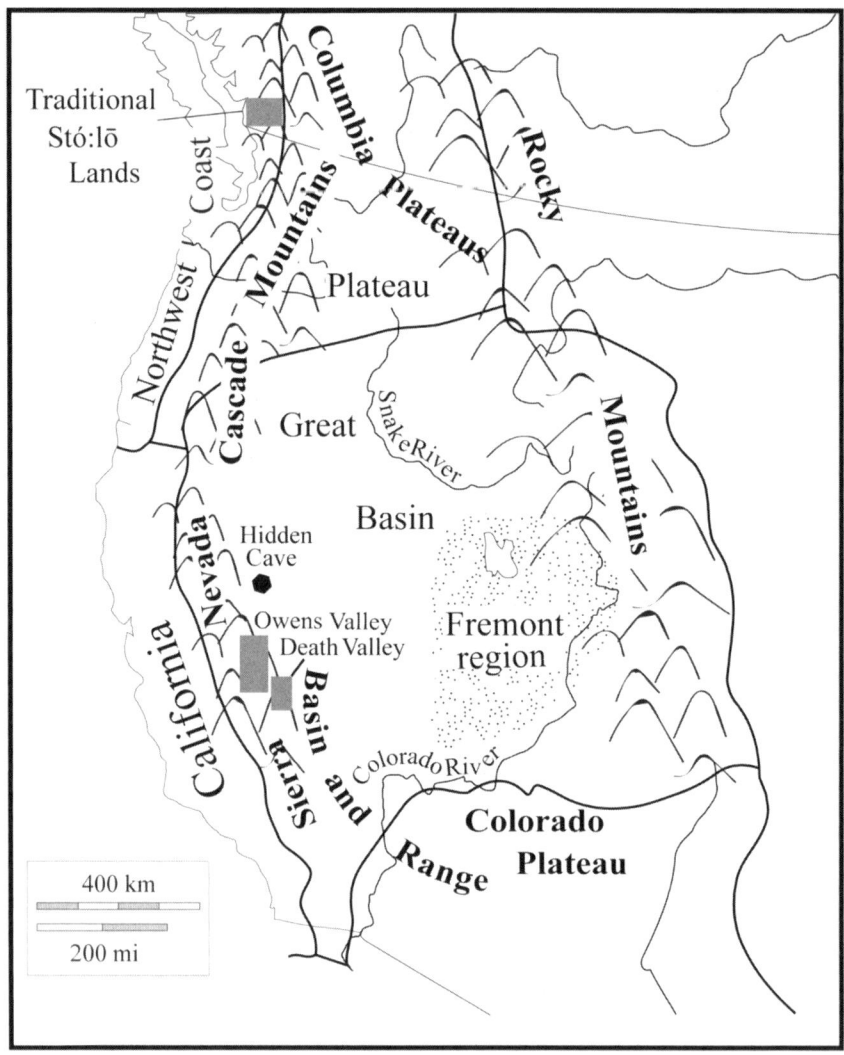

Figure 5.1. Map showing locations of culture areas and sites discussed in Chapter 5.

of the biomass is locked up in the structural tissue of woody plants, which has no food value for humans. The highly reticulate coastlines of its central area offer both high productivity in the form of marine and estuarine fish and invertebrates and considerable habitat diversity (Campbell and Butler 2011). The combination of patchiness, habitat diversity, and high productivity provides important clues to understanding Northwest Coast food production systems and their near-invisibility to Western observers.

Management of Plant Resources

Food production in the Pacific Northwest involved numerous species of plants, each with a distinctive set of management strategies and technologies. Turner and Peacock (2005:112) identify six categories of activities that have positive ecological effects on the targeted plants: selective harvesting and replanting, digging and tilling, tending and weeding, sowing and transplanting, pruning and coppicing, and landscape burning. These practices increase the production of useful parts (such as fruits, underground storage organs, or fibers) and help to protect economically important species from predation and competition. At the habitat and community levels, human interventions helped to increase the extent of edge areas between habitats (which tend to be highly productive), creating a mosaic of vegetation types. These efforts altered and in some cases even created new communities and ecosystems (Chambers and Turner 2011; Turner and Peacock 2005). They were employed in many ecological settings: low-elevation prairies and meadows, rainshadow forest (dominated by Douglas fir), coastal rainforest, freshwater wetlands, and tidal wetlands (Chambers and Turner 2011). Additionally, cultural rules regarding ownership and prescribed methods of harvest helped to prevent overexploitation. Considered collectively, plant management strategies improved both quality and quantity of resources, concentrated them in space, and ensured continuing productivity (Deur and Turner 2005).

Most of these practices target perennial plants, which adapt to seasonality by going into a dormant state rather than producing seeds and then dying after a year's growth. The fact that so many culturally important plants of the Pacific Northwest were perennials partially explains the failure of Westerners to recognize the existence of food production there; European economies were based primarily on cereal crops propagated by seed (Turner and Peacock 2005). Although archaeobotanical evidence is limited by the poor preservation potential of underground plant parts, early observations and linguistic reconstructions strongly suggest that the practices employed to enhance the yields and reliability of perennials predate European contact.

Many herbaceous perennials have underground storage organs such as roots, corms, and bulbs that allow these plants (collectively called geophytes) to store energy during the dormant period that will fuel the renewal of active

Figure 5.2. Common camas (*Camassia quamash*) flower (a) and bulbs (b). Photo credits: (a) William & Wilma Follette. *Western Wetland Flora: Field Office Guide to Plant Species*, 1992. West Region, Sacramento. Provided by USDA NRCS Wetland Science Institute. (b) T. Abe Lloyd.

growth when conditions improve. These storage organs are dense with food energy in the form of complex carbohydrates that can be broken down and made bioavailable by pit cooking (Chambers and Turner 2011; Wrangham et al. 1999). Some of the most important geophytes that were tended and harvested in the Pacific Northwest include springbank clover, camas (Figure 5.2), northern rice-root, fritillary, wild onion, wapato, chocolate lily, and yellow glacier lily. The act of harvesting underground parts of these plants aerates the soil in a manner beneficial to their growth. Harvesters left behind smaller-sized bulbs to continue growing, transplanted and replanted individuals to maximize growth, and removed competitors (Chambers and Turner 2011; Turner and Peacock 2005). On the coast, plants such as springbank clover, silverweed, rice-root, and lupine were grown in tidal marshes in garden plots that were demarcated and carefully tended. These sites were revisited on a seasonal basis to harvest geophytes in quantity for feasts, family gatherings, and storage (Turner and Peacock 2005). Such estuarine gardens often required considerable labor investment to create mounds of soil and build rock enclosures for soil retention. Remnants of these root gardens persist in the form of anthropogenic sediments that date to the precontact period (Deur 2005).

Drier inland areas were burned to maintain prairie vegetation as habitat for camas and other geophytes. There is archaeological evidence for harvesting and processing of camas as early as 11,000 years ago in the Willamette Valley of western Oregon, although it may have been practiced only sporadically until thousands of years later (Ames 2005). Where camas was harvested, periodic burning was required to maintain its preferred habitat. For this reason, the many examples of prehistoric camas-roasting pits from the Pacific Northwest can be taken as indirect evidence of intentional burning. Other evidence of prehistoric burning is the presence of remnant Garry oak meadows on Vancouver Island and in the Willamette Valley, where conditions are too wet for oaks to grow naturally and forest fires are rare (Lepofsky et al. 2005). Camas has also extended its range beyond its oak meadow habitat with human assistance that probably included prescribed burning. These anthropogenic prairies also enhanced local populations of other key resources such as deer, elk, edible berries, and nettles (Weiser and Lepofsky 2009).

Fruit-producing perennial trees and shrubs were key resources for peoples of the Pacific Northwest, so it is not surprising that they were tended as well. Like root gardens, berry patches were owned by the families and individuals who cared for them (Turner and Peacock 2005). Interviews with Stó:lō elders of the upper Fraser Valley of southwestern British Columbia indicate that prescribed fires were set in high-elevation areas in order to increase the size and quantity of blueberry fruits and limit pests and competition from other plants (Lepofsky et al. 2005). These practices are probably quite ancient, although their archaeological traces may be elusive. Lepofsky and colleagues (2005) relocated traditional high-elevation berry patches in traditional Stó:lō lands (Figure 5.1) but were unable to find a charcoal record of the activity in sediment cores.

Maritime Management

Evidence from both ethnography and archaeology have documented maritime management in the form of conservation measures (e.g., catch limits, harvesting decisions), habitat enhancement, and transplantation of fish. One such strategy is the construction of clam "gardens," boulder walls that create terraces to expand the area of optimal habitat for edible bivalves. Rock terraces also created a reef-like habitat attractive to other marine life. In British Columbia, Lepofsky and colleagues (2015) documented ancient clam gardens in several locations, with the earliest radiocarbon date indicating wall maintenance 1,000 years before present. Although clam gardens vary considerably in their setting and form, they all reveal careful planning to match wall height to the optimum tidal height for clam habitat.

Land Tenure

Aspects of land tenure and settlement in the Pacific Northwest are similar in many ways to those of agriculturalists. Whereas mobile foragers recognize rights of access to land and resources, these rights are typically fluid and negotiable. In contrast, subsistence agriculturalists recognize ownership of land by individuals, families, kin groups, or institutions (Bowles and Choi 2013; Woodburn 1982). A similar pattern is found on the Northwest Coast, where managed resource patches were owned by families or sometimes individuals. Access to clam beds was controlled by family heads (Lepofsky et al. 2015). "Root gardens" in tidal marshes and upland berry patches were owned by chiefs and families, and often the plots were marked in some way (Turner and Peacock 2005). These rights were widely recognized and respected within communities. It was understood that investment of labor in a resource patch served to establish a claim to ownership, or at least use-rights. Ownership brought with it responsibilities as well, and elite lineages played an important role in monitoring and controlling access to key resources.

Summary

Ethnographic and ethnohistoric evidence for food production in the Pacific Northwest is abundant for both coastal and inland areas. Naturally occurring concentrations of berry bushes, clams, and underground storage organs were enhanced by human interventions that ranged from subtle (clearing away competing vegetation or incidental soil disturbance) to carefully designed (terraced clam gardens, estuarine root gardens). Judicious use of controlled burning increased production of useful parts of plants and manipulated ecological succession to favor the growth of targeted species. Taking these practices into account, many groups of the Pacific Northwest meet all or nearly all of the criteria for agriculture used in this book, with the exception of the use of domesticates. However, we cannot be certain that domestication of geophytes would have resulted in recognizable changes in morphology, as is often the case for seed-producing crops (Smith 2005b). There is little justification for failing to acknowledge that Northwest Coast and Interior Plateau peoples were food producers (Smith 2005b).

The Great Basin: Flexible Food Production in a Marginal Environment

The Great Basin of the American West is, in addition to the Northwest Coast, a region that exemplifies low-level food production by groups generally categorized as hunter-gatherers (Smith 2005b). Much of this region consists of cold desert and high-altitude grasslands that are poorly suited for maize agriculture. However, Great Basin groups did practice management of selected populations and habitats through activities such as burning, clearing, pruning, coppicing, tilling, transplanting, and even some cultivation and irrigation (Fowler 2000). These and other forms of plant husbandry seem to have been at the center of a relatively stable subsistence economy rather than representing a temporary period of transition. Even on the southern edge of the Great Basin, where forager-farmer interactions led to the eventual adoption of maize, agriculture remained part of a highly flexible and diverse subsistence system. It supplemented, but did not replace, hunting and plant food collecting. These Great Basin maize farmers, known to archaeologists as the Fremont, have figured prominently in discussions of the utility of agriculture in the region (Barlow 2002; Madsen 1998).

Julian Steward, a central figure in mid-twentieth century American anthropology, used his ethnographic research among the Owens Valley Paiute of southeastern California (Steward 1933) as a springboard for theory-building. For Steward, the seed-collecting focus of this group emerged from the application of a particular, historically situated technology to the unpredictable and patchily distributed resources of the Great Basin (Morgan and Bettinger 2012). Later ethnographic work by Fowler (2000), Fowler and Rhode (2011), and Anderson (2005) has greatly enriched understanding of Great Basin plant husbandry by viewing it as an adaptive system grounded in traditional ecological knowledge. The archaeological record of food production is not as rich as the ethnographic one, but it has yielded some information about the time depth of practices reported by Steward and his successors. In addition, archaeologists have been interested in the adoption of maize by the Fremont as an adjunct to, rather than a replacement for, foraging. Together, these sources provide a basis for discussing the Great Basin as a nonagricultural region with a history of food production.

The Great Basin as Human Habitat

Generally the Great Basin refers to the area between the Sierra Nevada and Cascade mountain ranges to the west and the Rocky Mountains to the east, bounded by the Colorado Plateau to the south and the Columbia and Snake River plains to the North (Morgan and Bettinger 2012). Hydrologically, the Great Basin is the interior drainage system that lies between the Rocky Mountains and the Sierra Nevada. The many Pleistocene lakes of the region have largely disappeared or dried up, in some cases leaving behind vast salt flats or diminished bodies of water (including the Great Salt Lake). Ethnographically, the Great Basin culture area includes Shoshone and Paiute groups that expanded out of the Great Basin hydrologic basin into the adjacent Great Plains and Columbia Plateau. Not surprisingly, the region encompasses considerable environmental diversity that includes deserts, high plateaus, and mountains (Fowler and Rhode 2011). Most of it is occupied by two deserts, the low-elevation Mojave Desert in the south and the higher-elevation Great Basin desert in the north. Desert vegetation is largely shrub-steppe that is often dominated by big sagebrush (*Artemisia tridentata*). This arid landscape is interrupted by wetlands but receives little precipitation, much of which falls in the form of snow at higher elevations. Agriculture in the Great Basin faces many of the same challenges as it does in the Greater Southwest, but with a shorter growing season and even greater contrasts in temperature and moisture between drylands and forested mountains (Fowler and Rhode 2011; Morgan and Bettinger 2012).

The Small Seed Complex

The small seed complex of the Great Basin is associated with specialized technologies for collecting and processing. In some cases, these technologies included methods of sowing seeds and tending them during the growing season. Steward reported that the Owens Valley Paiute planted and irrigated flat sedge, bluedicks, and possibly spikerush (Steward 1933). They dug ditches that directed water to these plots from the mountain-fed streams of the Sierra Nevada. An irrigation "boss" was in charge of planning and supervising construction of feeder ditches and brush dams. Steward also saw the broadcast sowing of seeds in central Nevada among western Shoshone groups. These seeds included chenopod, Indian ricegrass, whitestem blazingstar, and an unidentified species. In this case, seeds were sown after burning (Fowler and Rhode 2011). Doebley (1984) compiled a more extensive list of 27 genera and 52 species of grasses that were used as food throughout the Southwest, including parts of the Great Basin. Cultivation of grasses was also practiced to the south on the Colorado River, an activity that required little investment of labor beyond broadcasting seeds on a muddy floodplain (Fowler and Rhode 2011).

It is not clear how ancient these practices are, but there is mounting evidence for the small seed complex at prehistoric sites in the Great Basin. Although the archaeological record does not provide definitive evidence for cultivation prior to circa 1840 (Fowler and Rhode 2011), it does indicate that the small seed complex has a deep history in the region. Material evidence of the small seed complex exists in the form of the abundant ground stone tools with which seeds were prepared and ceramic vessels in which they were cooked (Eerkens 2004). Archaeobotanical evidence is more informative about the plant taxa involved in the complex. Two components at the Sunga'va site in the Owens Valley dating between 1350 and 650 BP yielded carbonized assemblages of seeds that included several taxa known to have been used ethnographically (Eerkens 2003). The later component had a wider range of taxa and a higher density of seeds overall; one of these samples contained hundreds of seeds of bulrush (*Juncus*) that may have been stored. Rhode (2003) examined paleofeces from Hidden Cave, Nevada (Figure 5.1), and found a similar emphasis on small seeds during two occupation periods, from 1600–1450 BC and AD 50–450. Ricegrass and goosefoot were particularly abundant, with seepweed, saltbush, and blazingstar also present but not abundant. Some types of seeds from Hidden Cave—namely threesquare and cosmopolitan bulrush—show evidence of roasting and milling prior to consumption.

Management of Perennials

Like the peoples of the Pacific Northwest, Great Basin groups routinely protected and maintained perennial plants that produced edible seeds or fruits. These forms of management are not readily recognized by observers who are more familiar with annual planting and harvesting of seed crops. For example, the Owens Valley Paiute irrigated "wild" tubers along with small seeds (Fowler 2000). Excavation of pit-cooking features along the shores of Owens Lake suggests that this practice goes back at least 1,000 years (Eerkens et al. 2008). In Death Valley, the Panamint pruned mesquite to increase production of pods and help to disperse the seeds by discarding them during processing to extract the pulp. They sometimes pruned another important food plant, piñon pine, by pinching the tips of branches and whipping the higher branches with poles. Other woody plants such as willow and sumac were coppiced (cut back severely) in order to produce long, straight stems suitable for basketry (Fowler and Rhode 2011).

Great Basin peoples often used fire to encourage growth of valued plants, in some cases altering entire communities (Fowler 2000). They greatly valued wild tobacco (*Nicotiana attenuata* and *N. quadrivalvis* var. *bigelovii*) and enhanced its growth by periodic burning. The Timbisha Shoshone of Death Valley used fire to encourage wild tobacco (*N. attenuata*) grown in five-acre plots, to clear areas for planting maize and beans, and to reduce cattail dominance in marshes in order to attract waterfowl and create favorable conditions for *Scirpus acutus*, a bulrush valued for basketry. Some woody seed producers, such as blazingstar, also benefited from periodic burning. Fire played an important role in rabbit drives and also created prime habitat for edible seeds, which could be harvested the following year (Fowler 2000). In general, controlled burning created a heterogeneous mosaic of habitats and probably increased biodiversity.

Maize and the Fremont

Despite the challenges involved, the people known to archaeologists as the Fremont developed a distinctive subsistence adaptation between AD 500 and 1500 that combined cultivation of southwestern crops with hunting and collecting of indigenous plants. The Fremont culture was once thought to have emerged from Ancestral Puebloan populations who moved into the Great Basin, bringing their crops with them. However, more recent assessments view it as a historically deep desert foraging adaptation to which maize and other crops were added (Morgan and Bettinger 2012). The abandonment of Fremont sites between AD 1150 and 1350 may be related to the expansion of Numic-speaking populations into the region (Bettinger

1993), but some researchers emphasize the impact of periodic droughts (Coltrain and Leavitt 2002).

Reliance on maize among the Fremont was highly variable across time and space. Although some individuals from Fremont residential sites have ^{13}C values indicating maize was a major dietary constituent, others lack this signature (Zeanah and Simms 1999). Reliance on maize fluctuated in response to variation in moisture; during dry periods, maize was largely abandoned and wild foods became dominant once again (Coltrain and Leavitt 2002). Barlow (2002, 2006) showed that investment in maize among the Fremont was only profitable (that is, offering a relatively high rate of return) when foraged foods were in short supply. In years when environmental conditions favored plant growth, intensive agriculture was unable to match the return rates afforded by wild foods. However, small-scale planting of maize in untended plots is much less costly than tending large fields, and in less productive years, adding it to the roster of subsistence activities would have improved return rates beyond those obtained by foraging alone. By varying investment in maize (and avoiding commitment to intensive agriculture), Fremont communities were able to maintain their mixed economy despite high interannual variability in growing conditions for crops and wild plants alike.

Summary

Food production in the Great Basin was one aspect of a broad-based subsistence adaptation to a particularly challenging environment. One element of this adaptation involved augmentation of the yields of small seed-producing plants by sowing them in locations where they could be irrigated. Small seeds were important economically despite their relatively low return rates, probably indicating that limited abundance and unpredictability of more valuable resources made a broad diet advantageous. In this context, intensification of small seed production was an effective way to increase overall foraging efficiency in a marginal habitat. The same is probably true of the management of perennials, such as piñon and geophytes, through practices such as pruning and burning. Great Basin peoples also found ways to reap the economic advantages of maize cultivation when wild resources were scarce without becoming dependent upon agricultural production.

Summary

This chapter was originally conceived as an exploration of persistent foragers—people who chose not to engage in agriculture despite opportunities to do so. Close examination of ethnographic, ethnohistoric, and archaeological evidence indicate that being nonagricultural encompasses a broad range of options, most of which involve forms of intensification (Ames 2005) that are very different from the cultivation of domesticated seed crops. These practices—tending, controlled burning, selective removal of competitors, soil enrichment—deserve recognition as forms of food production. This characteristic unites village farmers, mobile desert dwellers, and coastal peoples with access to abundant and reliable marine and tidal resources. This broad definition of "food production" (Smith 2001, 2005b) acknowledges both the pervasiveness of intensification practices in North America and the distinctiveness of agriculture.

6

A World of Difference: Food Production in Postcontact North America

Columbus's landing on Hispaniola in 1492 has long been celebrated in the United States as marking a "discovery" that made America possible. However, celebration of Columbus Day is at odds with growing public awareness of the devastation wrought by Europeans in the New World. For wealth- and land-hungry European nations, this unknown territory held opportunities for colonial expansion and the capture of valuable natural resources. Native peoples were useful to these powers when they had a role to play as guides, forced labor, and hunters who could provide furs and skins for the transatlantic trade. But as the frontier of European settlement advanced, the presence of Native peoples and communities became inconvenient, leading to episodes of mass forced removal and armed conflict. It is this history that makes Columbus Day a day of mourning rather than celebration.

While attempted genocide, enslavement, removal, epidemic disease, and racism are no cause for celebration, the date itself commemorates one of the most significant events in world history. The first contact between Europeans and Native peoples in the Americas initiated a cascade of responses. Infectious Old World diseases spread rapidly and caused high mortality among indigenous populations that had no immunological experience with pathogens that had long been endemic in Europe. The survivors, though reduced in number, were frequently able to engage with the emerging colonial economies on their own terms, dictating which trade goods they were willing to accept and playing off European powers against each other as they vied for key alliances.

The initial and early periods of European contact also brought opportunities in the form of new technologies and goods, including newly introduced crops, domestic animals, and metal tools. In coastal areas, these opportunities involved direct interactions with colonists and in some cases subjugation or at least coercion by European powers (as was the case with the Spanish mission system). Interior groups often had access to European goods long before they had European neighbors, sometimes receiving these imports from Native middlemen rather than White traders. These groups were able to exercise considerable autonomy in their responses to European goods and tended to be highly selective, avoiding innovations that would require major subsistence reorganization (Gremillion 1993a). Some crops introduced by Europeans were readily adopted and integrated into traditional subsistence systems. In contrast, Old World agricultural technology such as draft animals and use of the plow made little impact initially. These patterns—which Old World crops and technologies were adopted, and why—are the subject of this chapter.

The effects that European contact had on Native food production were complex and varied. New biota appeared in the form of weeds as well as crops, and more subtle effects on landscapes were wrought by depopulation due to epidemic diseases. Some of the crops introduced by Europeans were well adapted to North American climates, whereas others failed to thrive. Agricultural tools and methods were generally slow to diffuse, except in the face of strong incentives (or disincentives), as when Native workers were expected to grow European crops to support colonists. Initial contact and colonization followed different historical pathways across the continent, with coastal groups sometimes undergoing rapid acculturation while interior groups reaped the benefits of indirect but profitable trade relations. The various European powers involved in conquest had contrasting goals that shaped their interactions with Native peoples; the French Jesuits, for example, attracted converts by showing tolerance and encouraging intermarriage, whereas Spanish missions practiced forced conversion and expected Native workers to support them economically (Lopinot 1986; Newsom and Gahr 2011). Native societies likewise had varying political and economic agendas that influenced their access to and attitudes toward novel items.

First I introduce a brief sketch of early explorations and attempts at colonization by various European powers before 1800. Then I present a broad-brush account of the introduction of crops from Eurasia, Africa, Mesoamerica, and South America, many of which were part of Spanish efforts to establish self-sufficient colonies anchored by missions (often Dominican or Franciscan) during the sixteenth century. Other crops became partly naturalized and made their own inroads into Native communities, or traveled in the form of seed as gifts or items for exchange. Domesticated animals, especially livestock, were sometimes viewed with suspicion on first encounter and often proved to be incompatible with traditional agriculture. Livestock had more success in the West than the East as the basis of cottage industries (hides and candle making) and ranching. More subtle than the appearance of novel organisms were some of the indirect effects of contact on Native agriculture, for example, the devastating reduction of populations through epidemic disease. These losses affected the ability to carry out communal tasks and transformed active fields into prime habitat for invasive Eurasian weeds. Following this sketch, I discuss in some detail the Spanish missions both east and west that encouraged or coerced the adoption of European-style agriculture and animal husbandry. I next turn to a consideration of broader patterns in the adoption (or nonadoption) of Old World crops, livestock, and food production technologies. A number of shared characteristics distinguish successful introductions from those that failed or had only a limited impact. Many of these characteristics are not inherent in the plants and animals themselves, but emerge from their interactions with new habitats and sociocultural systems. Social learning has an important role to play in the fate of innovations because it is often a key source of the knowledge required to employ them effectively. Cultural knowledge, inexpensively acquired through observation and instruction, reduces uncertainty and limits the costs of painstaking and time-consuming learning by trial and error.

Crop Introductions, Livestock, and a Changing Landscape

Old World plants and animals entered North America via numerous routes and mechanisms, many of them unintentional or incidental to the primary goals of the European powers. Monarchs needed new sources of wealth to pay for growing armies and administrative infrastructure in the context of trade competition with the East. Portugal and Spain sought gold, silk, spices, slaves, sugar, and a direct route to the Orient (Trigger et al. 1996). England and France strove to establish trade routes for access to animal products (furs in the north, deerskins in the south) and to open up the continent for colonization. All these activities brought Native people into contact with Europeans and Old World domesticates, whether through exchange of goods or gifts, forced labor, or in the domestic realm as interethnic marriages became increasingly common.

European Exploration and Colonization to 1800

Initial contacts between Native peoples and Europeans were limited to areas accessible by water. Columbus's landing on Hispaniola is the one that has greatest resonance in today's popular culture, but parts of the North American mainland also attracted the interest of European nations. Several nations of Atlantic Europe had by the early 1500s discovered the rich fishing grounds of the North Atlantic and established trade relations with Native communities. Verrazano and Cartier were the first to contact indigenous groups in the Northeast, in expeditions of 1524 and 1534–1541, respectively (Turgeon 1998). Trade with locals had become well established by the time of the earliest Jesuit narratives of the early seventeenth century.

The primary motivation for most of the visits to the Northeast by Europeans was exploitation of the rich fishing banks of the North Atlantic, which drew French, Spanish, Portuguese, and English cod fishers as well as Basque whalers. This trade introduced useful and decorative items, often made of metal or glass (Turgeon 1998), but these commodities seem not to have included foodstuffs, animals, or plants. On the other hand, Basque crews ate and drank with their trade partners, some of whom were able to understand French, English, Gascon, and Basque (Axtell 1988), and exchanged bread and cider for labor (Turgeon 1998).

To the south, the Spanish mounted expeditions into the interior Southeast and established mission settlements along the coasts of South Carolina, Georgia, and Florida, including Santa Elena in 1566 and St. Augustine in 1565 (Trigger et al. 1996). While some of the missions persisted into the seventeenth century, Spanish parties exploring the interior constructed forts but failed to establish permanent colonies. French colonies at Charles Fort in 1562 and Fort Caroline in 1564 (Figure 6.1) also failed, as did the Roanoke colony established by the English in 1585. Before 1670, when the English established Charles Towne (modern Charleston, South Carolina), however, most of the contacts between the English and interior groups were indirect ones conducted through Native trade intermediaries (Newsom and Gahr 2011; Waselkov 1989). By this time the English were firmly established along the mid-Atlantic coast and had developed an extensive trade network into the interior. The French, concerned about competition with

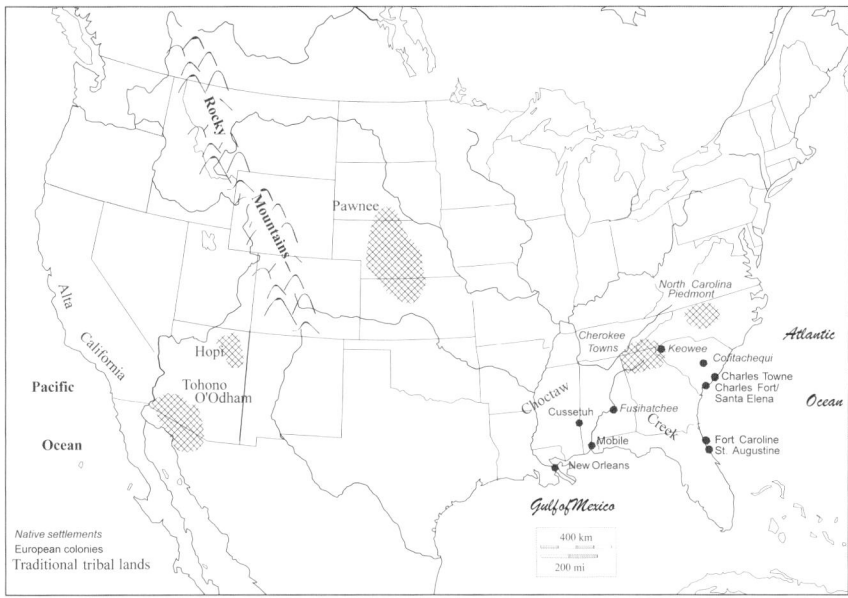

Figure 6.1. Map showing sites and tribal lands discussed in Chapter 6.

the English, were active in the lower Mississippi Valley and established colonies at Mobile (1702) and New Orleans (1718) at the start of the eighteenth century (Gremillion 2002b).

The Spanish began exploring the Pacific coastline following the conquest of Mexico (Trigger et al. 1996), entering the Baja Peninsula in the 1530s and the Colorado River delta in 1540. Many of the colonists came from the mining and agricultural towns of northern Mexico (Trigger et al. 1996). The Colorado River expedition of Alarcón went as far as its confluence with the Gila River (Lopinot 1986). Spanish missions were established in the eighteenth century in Alta California and the Southwest (Pavao-Zuckerman and Reitz 2011). Northern California had some trade contacts with Europeans, for example with Sir Francis Drake's crew in 1579, and obtained European goods from salvage of shipwrecks (Trigger et al. 1996). Peoples of the Northwest Coast had contact with Russian colonists during the late seventeenth century, but the Interior Plateau received trade goods indirectly until the arrival of traders and colonists in the wake of the Lewis and Clark expedition of 1805–1806 (Newsom and Gahr 2011). Colonization of the Plains, Great Basin, and Interior Plateaus was largely a nineteenth-century phenomenon (Fisher 1996; Fowler et al. 1996; Green et al. 1996).

The Postcontact Landscape

These earliest contacts had ecological and biological impacts that far outstripped the immediate consequences of face-to-face encounters. In many regions, initial contacts were trade-based and friendly but infrequent and informal. Even where colonization efforts followed rapidly on the heels of exploratory journeys, as was the case with Spain, they exposed large Native populations to contact with Europeans only indirectly. However, indirect pathways were sufficient for the spread of microorganisms, including those responsible for devastating Old World infectious diseases. These foreign pathogens moved rapidly through Native populations that lacked immunological experience with them, and mortality rates were high (Crosby 1976, 1986; Ramenovsky 1987). Severe depopulation during the Protohistoric period (after 1492 but before direct contact with Europeans) had widespread and pervasive effects on Native economies and social systems that had been left to function with a greatly diminished labor pool.

The effects of depopulation on agricultural practice are difficult to identify through documentary and archaeological evidence, but frequent observations in the East by European travelers of "old fields" probably refer to reduction in the amount of land being farmed by Native groups. An example comes from the account of John Bartram, a naturalist who trekked through the Southeast between 1773 and 1776 (Bartram 1928). He observed near the Cherokee town of Keowee many of these abandoned old fields, testament to the losses of community members to epidemics and the scorched earth campaigns waged by Whites (Hatley 1989). Much earlier, de Soto's army passed empty towns grown up in weeds near the town of Cofitachequi (Figure 6.1), also perhaps a sign of depopulation in the wake of disease (Trigger et al. 1996). The survivors, reduced in number, encountered problems coordinating usual subsistence activities, such as maintenance of communal fields that were located some distance away from towns. Men increased burning of woodlands in pursuit of deerskins for the profitable trade, which may have negatively impacted the gathering activities of women (Hatley 1989). No doubt similar reorganization efforts were taking place across the Southeast as people tried to maintain agricultural production with a greatly reduced labor pool.

As depopulation lessened the anthropogenic impacts on North American landscapes, introduced weeds began to compete with, and in some cases replace, the indigenous plants adapted to disturbed habitats. In Cherokee country, European-introduced weeds began to colonize the open parklands near the village of Keowee (Hatley 1989). Some of these Eurasian weeds had coevolved with livestock and were dispersed along with animal husbandry (Crosby 1986; Gremillion 2002b). For example, curly dock and the common dandelion probably entered North America both as cattle feed and as edible greens (Newsom and Gahr 2011). In the Southeast, goosegrass and prickly mallow were present at Spanish mission sites of the sixteenth century, and by the mid-eighteenth century, goosegrass was also found at Fusihatchee village in Creek country (Gremillion 1995b, 2002b). Documentary records affirm that dandelion, shepherd's purse, and plantain were all present in the Southeast during the seventeenth century; in fact, plantain was so closely associated with the English that it was known as "Englishman's foot." Contemporary observers remarked that plantain was closely associated with cattle dung, which facilitated its spread (Crosby 1986; Gremillion 2002b).

Old World Crops

On both coasts, missionization efforts by Spain brought numerous Iberian crops to North America in an attempt to set up self-sufficient communities. For the most part, other European powers lagged behind and in many cases relied on indigenous crops and support from the homeland. Prospective colonists from England were tempted with fertile farmland, some of which was intended for the production of non-native crops—rice, indigo, and cotton. Subsistence farmers also grew in number, and they brought an array of crops that were familiar to them. However, few of these crops spread to Native communities that were not in direct and sustained contact with Euro-Americans.

EASTERN UNITED STATES. In the East, sixteenth- and seventeenth-century Spanish settlements were supplied with a suite of Iberian crops that included staples such as wheat as well as familiar fruit trees, including oranges, peaches, and apricots (Table 6.1; Newsom and Gahr 2011). The Spanish also introduced numerous crops from elsewhere in New Spain, including lima beans and tomatoes from South America and butternut squash from Mexico. British introductions included rice, wheat, barley, oats, rye, peas, flax, hemp, indigo, tobacco, and cotton. In the case of Spanish missions (discussed further below), Native peoples were persuaded or forced to grow Old World crops for the support of colonists. However, many of the

Table 6.1. Selected Crop Plants Introduced to North America by Europeans[a].

Common Name	Scientific Name	Geographic Origin
Cereals		
African rice	*Oryza glaberrima*	West Africa
Asian rice	*Oryza sativa*	South China/India
Barley	*Hordeum vulgare*	Southwest Asia
Oats	*Avena sativa*	Northern Europe
Rye	*Secale cereale*	Anatolian Plateau/Northern Europe
Sorghum	*Sorghum bicolor*	Africa
Wheat	*Triticum aestivum*	Transcaucasia
Legumes		
Chickpea	*Cicer arietinum*	Southwest Asia
Cowpea	*Vigna unguiculata*	West, Central Africa
English pea	*Pisum sativum*	Mediterranean
Fava bean	*Vicia faba*	North Africa/Southwest Asia
Hyacinth bean	*Lablab purpureus*	Africa, India, East Asia
Roots and tubers		
Andean potato	*Solanum tuberosum*	South America
Beet	*Beta vulgaris*	Mediterranean/Western Europe
Carrot	*Daucus carota* ssp. *sativus*	Eurasia
Sweet potato	*Ipomoea batatas*	South America, Caribbean
Yam	*Dioscorea* spp.	Africa
Fleshy fruits		
Apricot	*Prunus armeniaca*	Turkey–Iran
Currant	*Ribes* spp.	Eurasia
Date palm	*Phoenix dactylifera*	Southwest Asia
Fig	*Ficus carica*	Mediterranean
Olive	*Olea europaea*	Mediterranean
Orange	*Citrus aurantium, C. sinensis*	Mediterranean
Peach	*Prunus persica*	China (west temperate)
Pomegranate	*Punica granatum*	Eurasia
Watermelon	*Citrullus lanatus*	Central and southern Africa
Wine grape	*Vitis vinifera*	Mediterranean
Vegetables		
Asparagus	*Asparagus officinalis*	Eurasia, North Africa
Butternut squash	*Cucurbita moschata*	Central, South America
Cabbage	*Brassica oleracea*	Western and southern Europe
Cauliflower	*Brassica oleracea* var. *botrytis*	Mediterranean
Curly dock	*Rumex* spp.	Eurasia
Onion	*Allium cepa*	Mediterranean
Tomato	*Solanum lycopersicum*	Central America
Winter squash	*Cucurbita maxima*	Central, South America
Fibers, dyes, flavorings, and medicines		
Chili, red pepper	*Capsicum* spp.	Central, South America
Cotton	*Gossypium* spp.	Africa, Asia Minor, India
Flax	*Linum usitatissimum*	Asia?
Hemp	*Cannabis sativa*	Central Asia
Indigo	*Indigofera* spp.	Various Eastern Hemisphere
Sweet basil	*Ocimum basilicum*	Tropical Eastern Hemisphere
Sugarcane	*Saccharum officinarum*	Southeast Asia
Tobacco, domestic	*Nicotiana rustica*	South and Central America

[a] Source: Newsom and Gahr (2011).

Iberian crops did poorly in the East, far from the long growing seasons and ample sun of Mediterranean Europe (Ruhl 1993). Some of the fruit trees became established at lower latitudes, especially in peninsular Florida. Spanish missions of the Gulf and Atlantic coasts have proven to be particularly rich contexts for the archaeological documentation of agriculture in these multiethnic communities (Scarry 1985, 1991; Scarry and Reitz 1990).

On the mid-Atlantic coast, English plantation owners developed rice culture supported by the labor and knowledge of enslaved Africans. Many were familiar with the native African cultivated rice *Oryza glaberrima,* and applied that knowledge to the Asian *Oryza sativa* (Carney 1993; Carney and Porcher 1993). In the Northeast and the Maritime Provinces of Canada, a number of crops were introduced as early as the early seventeenth century by fur traders, missionaries, and European settlers. Watermelons, peas, peaches, and apples were taken up by the Huron and Iroquois. Some root crops did well in the Northeast; Andean potatoes were introduced in 1621, and turnips were grown at a Moravian mission on the Labrador coast (Newsom and Gahr 2011).

By the time English colonies had become well established on the Atlantic Coastal Plain in the seventeenth century, Native communities and populations had been greatly reduced by Old World pathogens and forced to reorganize and resettle. Relatively little is known about how these coastal groups responded to the presence of novel plant foods, although there is evidence that the peach and the cowpea were acquired either directly from the Spanish, from other Native groups, or from naturalized populations (Clifford 2012). In the interior Southeast, Native villages formed trade relationships with Europeans indirectly, by way of Native middlemen who profited greatly from this role (Gremillion 1993a; Waselkov 1989). For roughly two centuries, these interior groups had opportunities to selectively incorporate trade goods into their cultural repertoire. Whether new crops traveled along these trade routes is not known; what is clear is that very few Old World crops penetrated the interior prior to European colonization in the mid-eighteenth century. The exceptions were the peach, which had rapidly become naturalized after the Spanish introduced it in the sixteenth century, and two African crops, cowpea and watermelon.

The fates of these three crop plants in the interior Southeast illuminate some of the decision criteria at work among recipient groups (Gremillion 1993a, 1996a). The peach is native to Asia and may have been introduced to the Americas as early as Columbus's second voyage. The Spanish brought it to the Southeast, probably early in the sixteenth century. Peaches were a frequent component of Spanish mission gardens along the Atlantic and Gulf coasts following the founding of St. Augustine and Santa Elena in the 1560s. In the interior, archaeological evidence in the form of charred peach pits becomes increasingly common after 1620 (Knight 1985:79; M. Smith 1987:125). Peach pits are common on Cherokee sites in the Appalachian uplands of eastern Tennessee (Chapman and Shea 1981) dating to the late eighteenth century. Unlike many Iberian crops, the peach did well in the Southeast and spread rapidly (Reitz and Scarry 1985:55; Ruhl 1993); so well, in fact, that the Tuscarora of eastern North Carolina claimed of at least one variety that "they had it growing amongst them, before any Europeans came to America" (Lawson 1967 [1709]:115). Assisting the rapid spread of this tree is its ability to germinate spontaneously from discarded seeds, after which it takes only three to five years to bear fruit (Sheldon 1978). Thus, even taking into account the propagandistic nature of John Lawson's account of the agricultural potential of North Carolina (Lindgren 1972), it does not stretch the imagination too much to accept that "Eating Peaches in our Orchards makes them come up so thick from the Kernel, that we are forced to take a great deal of Care to weed them out; otherwise they make our land a wilderness of Peach-Trees" (Lawson 1967 [1709]:115). The weedy ecological tendencies of the peach assisted their spread throughout the Southeast amongst groups that had trade relations (however indirect) with Europeans.

The seeds of the watermelon are far less likely than peach pits to be preserved by charring, which is the primary mode of preservation for archaeological plant remains in the hot and humid climate of the region. Archaeological evidence is sparse, consisting of one seed from each of two late seventeenth-century North Carolina Piedmont villages (although other instances have been reported from seventeenth- and eighteenth-century sites in the Midwest; Blake 1981; Gremillion 1993a). However, documentary evidence indicates that watermelon was present in Spanish colonies of the Atlantic coast by the final quarter of the sixteenth century (Blake 1981; Reitz and Scarry 1985). As early as 1597, it was being grown in the interior of what is now Georgia (Blake 1981). The response of Native communities to the watermelon was universally positive; according to the Lincecum Manuscript, a Choctaw folk narrative, "They were so much delighted with the watermelon that they saved every seed and cleared large plots of ground for the next year" (Campbell 1959:16).

The cowpea or black-eyed pea is an infrequent find in the Southeast, but was recovered in quantity (over 300 whole seeds and an equal number of cotyledons, the embryonic leaves that make up half of each whole "pea") within a burned structure at the Upper Creek village of Fusihatchee. This find places the cowpea in central Alabama by 1670, well before its appearance at Cherokee sites (Chapman and Shea 1981; Gremillion 1995b; Waselkov 1989). Cowpeas were most likely introduced here through middleman trade with Spanish coastal colonies (Waselkov 1989). This inference is supported by the label "Apalachean bean" used by Le Page du Pratz for a brown legume with a black ring around the hilum or "eye" in reference to the Apalachees, a group that brokered trade between Spanish sources and interior groups (Le Page du Pratz 1972 [1758]:7). Although not reported archaeologically from Spanish mission sites, cowpeas may originally have been misidentified as the common bean.

Introduced crops did not make serious inroads into the southeastern interior until after the end of the Yamasee War in 1717. This conflict suppressed a rebellion of Native trade middlemen, and the treaty that marked its end reestablished friendly relations between the English and their Native trading partners. However, this agreement reconfigured the distribution chain to rely on English traders who took up residence in Native villages, frequently cementing their ties to the community by marrying local women (Braund 1993:34–36). By the latter half of the eighteenth century, the Upper Creeks of Alabama had adopted sweet potatoes, peas, wheat, and rice (Gremillion 1993a, 1995b; Newsom and Gahr 2011), and Cherokee women were growing sweet potatoes in their infield gardens (Hatley 1989). At the Lower Creek town of Cussetuh (1715–1836), traditional foods such as maize continued to dominate Native diets, making up 86% of the assemblage of cultivated grains, which also included rice, wheat, and potatoes (Foster 2010). Documentary sources indicate that during the eighteenth century, the Choctaws grew several Eurasian garden vegetables in addition to peach, watermelon, and the African grain crops guinea corn and sorghum (White 1983:19–20).

These introductions notwithstanding, there was considerable resistance to adoption of European agricultural technology. The Creeks are reported to have deliberately rejected use of the plow in order to prevent traders from producing their own food (Braund 1993:75), and in general Native women were a conservative influence, continuing to process food and make pottery using traditional methods despite the availability of European goods (Braund 1993:30–31).

WESTERN UNITED STATES. The Greater Southwest has a substantial record of the introduction of crops in conjunction with the establishment of missions, which played a key role in this process. However, these crops were not universally accepted by Native farmers, and those that were difficult to grow were largely ignored. Wheat, for example, yielded less and needed more water than maize and was grown primarily to support missionaries (Lopinot 1986). An exception to this generalization comes from southern Arizona, where wheat fit well into the traditional irrigation schedule for crops along the Gila River (Newsom and Gahr 2011). Another factor contributing to the rapid adoption of wheat is that it became ripe in May, when many other foods were scarce (Lopinot 1986).

Crops that did find a place in Native agricultural systems were most often additions to, rather than replacements for, traditional ones. Many cucurbits (gourds, squashes, and melons) were adopted readily, as were some of the fruit trees introduced by the Spanish (Lopinot 1986). By 1690, the Hopis were growing peach, apricot, plum, and cherry trees in addition to chilies, onions, watermelons, and cantaloupes (Trigger et al. 1996). Although frequently introduced as mission crops, some (including watermelons and peaches) spread rapidly beyond the missions' sphere of influence by moving directly between Native communities. The importance of these indirect mechanisms of introduction is illustrated by the fact that the Tohono O'Odham of southern Arizona considered cowpeas, chickpeas, and lentils to be native, not introduced (Newsom and Gahr 2011).

More remote groups in the interior West, including the Great Basin, did not confront Eurasian crops prior to the nineteenth century (Newsom and Gahr 2011). However, indirect trade allowed the Northwest Coast peoples to acquire potatoes and turnips, which they grew in addition to the indigenous geophytes (Trigger et al. 1996). On the Great Plains, Old World plants, including watermelons, cantaloupes, peaches, and English peas, made inroads during the early eighteenth century. The southern Plains region traded with the southwestern Pueblos and presumably had access to crops introduced by the Spanish, although archaeobotanical evidence is lacking. In general, adoption patterns on the Plains resembled those seen in other North American regions in consisting of selective additions to the roster of indigenous crops (Adair and Drass 2011). The Pawnees, maize farmers, and hunters of the central Plains, adopted the watermelon and possibly the Andean potato as crops after contact (White 1983:159).

Exotic New World Crops

Some introduced crops originating in Mesoamerica or South America arrived in North America indirectly, by a circuitous route following an initial dispersal to Europe. Spanish conquest of Mexico and Andean South America facilitated the spread of a number of crops that were still absent from North America in precolumbian times. For the most part these were not staples, but specialty crops used ceremonially or medicinally or primarily for flavor and seasoning. The English introduced a new type of tobacco, *Nicotiana tabacum*, after smoking and snuff-dipping had become popular fads at home. Native North Americans were already familiar with some native tobaccos (in the western United States) and *N. rustica*, the high-nicotine species that diffused in precolumbian times ultimately from South America (Dunavan and Jones 2011; Newsom and Gahr 2011). Another crop obtained indirectly by way of Spanish colonization of South America was the Andean potato, which was quickly adopted by Northwest Coast groups. It may have been introduced by Russian colonists in the early nineteenth centuries to the Northwest Coast, but seems to have arrived in eastern North America by way of Europe as early as 1621 (Sauer 1994:150–151).

Livestock

Chickens had some parallels with the turkeys domesticated in the Greater Southwest, but for the most part the relationship between humans and mammals was that of predator and prey. In contrast to crops, hoofed livestock had no precedent in Native subsistence economies. At first, horses, cattle, sheep, and goats were regarded with suspicion by many Native groups, and in some contexts they actually interfered with production of food crops. However, in others, domesticated animals presented new and profitable economic opportunities.

Livestock were more successful in the Southwest than they were in the Southeast, in part due to environmental factors. Sheep and goats did well in arid and semiarid habitats, converting inedible plant material into meat and thus presenting a profitable alternative to the challenges of drylands farming. Livestock theft, pursued on horseback, was also a potentially profitable venture that some Native groups embraced enthusiastically. In the Southeast, domestic animals had very limited impact on Native diets until the nineteenth century, although horses quickly found use transporting trade goods. Perhaps the peltry trade that dominated Native–European economic relations absorbed much of the labor that could have been devoted to livestock (Pavao-Zuckerman and Reitz 2011).

EASTERN UNITED STATES. The de Soto expedition of 1540 introduced hundreds each of cattle, sheep, horses, pigs, and goats into the Southeast (Pavao-Zuckerman and Reitz 2011:581). Some of the pigs escaped and established feral populations, but not surprisingly, none of the communities on de Soto's path adopted animal husbandry. The integration of livestock into Native economies did not begin until the establishment of missions between 1565 (the founding of St. Augustine) and 1763 along the Gulf and Atlantic coasts (Pavao-Zuckerman and Reitz 2011:581). Missions played a major role in this process because the Spanish colonial enterprise involved not only religious conversion but also the transformation of converts into farmers who could support colonial efforts. Despite this aim, however, zooarchaeological assemblages from southeastern missions reveal that consumption of meat from domestic animals was more infrequent than historical documents suggest (Pavao-Zuckerman and Reitz 2011; Reitz 1991; Reitz and Scarry 1985).

Outside the sphere of influence of the missions, Native villages in the interior Southeast were slow to adopt livestock, which did not become common there until the nineteenth century (Pavao-Zuckerman and Reitz 2011). Among the Lower Creeks, who had especially close contacts with European neighbors and trade agents, some residents began to raise cattle and pigs after 1800. However, the evidence of introduced livestock is highly variable between Lower Creek sites that have been investigated archaeologically (Foster 2009). In Upper Creek country (central Alabama) and the villages of the North Carolina Piedmont, archaeological evidence of livestock is sparse prior to the nineteenth century. For example, the seventeenth-century component at Fusihatchee village had bones representing one pig, one possible cow, and two chickens out of a total of 127 individuals. The eighteenth-century component was similar, with two pigs, five (possible) cows, and four chickens, but also included an equid (probably a horse). The Fredricks site in North Carolina, occupied between 1680 and 1710, yielded only a single element each of pig and horse. The Cherokees of North Carolina and Tennessee made use of horses to facilitate trade starting in the mid-eighteenth century (Pavao-Zuckerman and Reitz 2011). However, Cherokee women disliked horses and other domestic animals because they interfered with garden crops (Hatley 1989). Villages even more remote from trade outposts, in the Appalachian highlands of Virginia, have so far produced no evidence of introduced livestock in the sixteenth and seventeenth centuries (Pavao-Zuckerman and Reitz 2011).

In the Northeast, English settlement during the seventeenth century and later was dependent upon livestock for manure, traction, meat, and dairy and wool production. This reliance on animal husbandry was perhaps the most significant difference between Native and Euro-American agriculture, and one that created conflict as grazing and foraging livestock damaged Native fields and their crops (Cronon 1983:128). Although in close proximity to European settlements, Native communities of New England had little incentive to keep livestock in the context of traditional farming systems.

The Great Plains present an unusual case of the rapid adoption of a European domesticate, the horse, independently of conversion to European mixed farming. However, Plains groups acquired horses to be more effective warriors and hunters rather than to mimic European farming styles. By the late seventeenth century, Native groups were using horses they obtained from the Spanish and southwestern Pueblos for hunting and transport (Fowler et al. 1996).

WESTERN UNITED STATES. In contrast to the situation in the East, in the western United States European livestock rapidly became economically important following their introduction. In many cases, Native people took up animal husbandry in the shadow of mission settlements and in their service (see below for further discussion of the mission system). However, autonomous groups such as the Navaho and Apache quickly adopted sheep and goats, as did some of the Pueblo groups. In the arid habitats of the Southwest, livestock (especially sheep and goats) were effective at converting inedible grasses into an edible meat supply (Pavao-Zuckerman and Reitz 2011). However, draft animals were not widely used until the nineteenth century.

The Spanish Mission System: Native Labor, Agriculture, and Colonial Expansion

The rapid incorporation of livestock as well as many Old World crops into Native economies in both the Southeast and the Southwest was closely tied to the Spanish crown's strategy of colonization in the New World. In both parts of the continent, religious orders established missions as part of the larger colonization strategy of the Spanish Empire in New Spain. The role of these missions, often supervised by Franciscan or Dominican friars, was not only to convert Native people to Christianity but also to create new agrarian communities that could support the colonial venture (Pavao-Zuckerman and Reitz 2011). In some respects, "conversion" to agriculture was no less essential to the mission enterprise than religious conversion (Newsom and Gahr 2011). Unlike the interior groups who maintained largely traditional subsistence economies while selectively incorporating European goods, coastal groups in mission areas were immediately pressured to produce agricultural products for the support of the mission. The missions demanded tithes and showed leniency to Native groups who were willing to cultivate European crops (Lopinot 1986), and labor was sometimes forced (Scarry 1991). Under these conditions of coerced acculturation, missionized Native groups rapidly adopted "Europeanized" diets (Trigger et al. 1996).

The Spanish entered the New World colonization effort with the aim of recreating the Iberian way of life, including its culinary and agricultural traditions (Ruhl 1993). Many of the crop introductions associated with Spanish missions reflect this goal. Wheat and the wine grape were culturally essential because of the role they played in the Eucharist as the raw material for wine and communion wafers (Newsom and Gahr 2011). Traditional Spanish cuisine relied on olive oil and bread as well as onions, garlic, radishes, lettuce, cucumbers, garbanzo beans (chickpeas), lentils, peas, and a variety of herbs and spices (Ruhl 1993). Several tree crops were introduced to mission settings, including peaches, apricots, and oranges. Unfortunately for the Spanish, many of the Iberian crops did poorly in the hot and humid subtropical climate of the Southeast and had to be abandoned (Ruhl 1993). In the arid Southwest and Alta California, many of the same crops did well, although not all of them were embraced by Native communities. Fruit trees such as the peach that required minimal care persisted, but grape and olive, which were more costly to plant, tend, harvest, and prepare for use, were less likely to be adopted (Lopinot 1986). Wheat was widely grown, and some Native groups found a place for it in their subsistence economies, although the bulk of this crop went to missions for use in bread and communion wafers. The basis of Native diets remained much the same, with maize, beans, and squash remaining dietary staples.

Many of the missions in Alta California, the first of which was established in 1769, focused on livestock production for meat, tallow, and hides (Pavao-Zuckerman and Reitz 2011). The Spanish established presidios and missions along the California coast in an effort to compete with Russian exploration and colonization. In the Southwest, missions among Pueblo groups introduced cattle, which became important as sources of dung for fuel as well as for meat. Missionaries seem to have consumed beef more frequently than did Native villagers, although some welcomed it as a replacement for bison that had once been obtained through trade with Plains groups. There was some variation in the dietary role of meat from domestic animals judging by the distribution and relative quantities of faunal remains in mission contexts. In Alta California, Native neophytes living in the missions ate large amounts of beef along with some wild game, and they adopted Spanish-style butchering methods. In contrast, Puebloan villagers who were associated with missions used traditional butchering methods, although they did consume some livestock (Pavao-Zuckerman and Reitz 2011). This comparison hints at the importance of social learning in the transfer of innovations.

Summary

For the first few centuries of European contact, during which most Native communities retained considerable autonomy, the incorporation of new animal and plant resources was highly selective. For agriculturalists, maize remained the primary crop. New ones were added, but only if they entailed little risk of adverse outcomes and could be incorporated easily into existing production systems. Crops that fit these criteria were often non-staples, such as fruit-producing trees and melons, and often had analogs among the suite of traditional crops that could serve as a source of relevant knowledge about how to grow and consume them. In the absence of such analogs, unfamiliar crops were probably more likely to be adopted where opportunities for social learning existed—in mission settings, or in the households of English traders with Creek or Cherokee wives. In their choices, Native peoples displayed both resilience in the face of rapid change and an openness to innovations that could enhance traditional foodways without disrupting or replacing them.

7

Synthesis

Food production in Native North America took many forms. A variety of adaptations is to be expected given the great topographic, biotic, and climatic diversity of the North American continent. Major environmental contrasts set constraints that remained relatively stable over time—for example, the aridity of much of the West and the short growing seasons of the North. Such variables shaped the complexes of crops that flourished in different regions and stimulated experimentation with intensification techniques designed to maintain or increase productivity. Nonetheless, it is clear that the historical development of food production is underdetermined by features of the wider environment in which human groups made a living. Observations about the natural and social worlds were grist for the mill of decision-making by individuals and groups about which foods to pursue, harvest, or enhance. The archaeological record is the residue of such decisions and (we assume) preserves a reliable record of major shifts in subsistence. We can expect to find evidence, for example, of the initial adoption of a crop plant, expansion of populations into new habitats, anthropogenic vegetation change, and even the process of domestication itself. Because of the importance of human decisions in explaining change, it makes sense to spell out some of the assumptions about human behavior that underwrite the subsequent discussion.

Evolutionary Ecology, Social Learning, and Food Production

The theoretical framework used here aims to understand historical changes in human behavior as emergent properties of decisions made by individuals in particular ecological settings. It draws on the field of evolutionary ecology, which seeks to explain adaptive strategies in environmental context

(Winterhalder and Smith 1992). In the case of humans, these strategies emerge from a highly flexible behavioral repertoire that permits rapid adjustment to changing conditions. Although this form of phenotypic adjustment is not unique, in the human case behavioral innovations (and the information that underwrites them) can be rapidly transmitted via social learning. This pathway of cultural transmission interacts with, and is parallel to, the biological one that transmits genetic information between generations (Richerson et al. 2010). Culture allows knowledge to accumulate within communities (creating bodies of traditional ecological knowledge, or TEK), while experimentation and individual learning provide ongoing refinement (Boyd et al. 2011; Henrich et al. 2016; Mathew and Perreault 2015). Much like the process of natural selection, cumulative social learning has a tendency to preserve practices that make their possessors more likely to survive and serve as cultural models for others (Henrich et al. 2016; Richerson and Boyd 2005).

The flexibility of this adaptation rests largely on the ability to process information rapidly enough to permit adjustment to changing conditions. Behavioral flexibility allows many such adjustments to take place during the lifetime of the organism, rather than exclusively through the weeding out of unsuccessful variants gradually across many generations. Humans support their extraordinary degree of behavioral flexibility by processing information about the biotic and social environments in which they live. These decisions represent the application of evolved cognitive biases, in the form of general rules of thumb, to specific survival problems (Boyd and Richerson 1985; Gigerenzer and Goldstein 1996). Many such biases probably exist, but to answer questions about the factors driving food production, there is justification for focusing on assessment of energetic payoffs. In evolutionary ecology, energy efficiency is often used as a proxy for fitness on the grounds that time saved can be reinvested in acquiring more energy or in other activities that promote survival and reproduction (Smith 1979). Therefore, subsistence behavior and the cultural patterns that result (and their archaeological traces) cannot be understood without examining their economic logic. Most archaeologists employ this assumption implicitly by referring to improved harvest yields, resource-rich habitats, and increases in resource abundance (Gremillion et al. 2014).

Another cognitive bias that can be confidently linked to survival and social success is risk assessment (Marston 2011; Winterhalder 2012; Winterhalder et al. 1999). Risk is probabilistic variance in outcome, and under most conditions the greatest likelihood of achieving some minimum payoff is ensured by reducing that variance—that is, by choosing the less risky of two strategies that yield the same average outcome (Winterhalder et al. 1999). Only when this minimum payoff is well above the expected average does the high-variance option offer a better chance of meeting the minimum. Otherwise, there is a strong incentive for people to try to limit risk by depending upon highly reliable food sources, maintaining networks of sharing and exchange, varying field locations, and storing surplus for future use. Such practices tighten up the frequency distribution of food availability to cluster around the mean, making a shortfall less likely under most conditions. Methods of reducing risk vary, but efforts to do so are expected to be pervasive and likely influential in subsistence decisions.

I assume that individual or household-level decisions based on these criteria shape the aggregate patterns that characterize cultures and communities. What makes subsistence patterns unique is the deployment of these species-typical modes of reasoning in different socioecological contexts. Resource abundance and density, spatial distribution, and the conditions that influence them are important determinants of economic choice. Rather than being a static backdrop, however, environmental parameters are constantly shifting, in many cases as a result of human activities. Whether they take the form of "ecological engineering" (Odling-Smee et al. 1996) or unintended effects, these impacts set up new environmental circumstances that influence decisions, creating a feedback loop in a process that has been labeled niche construction (Kendal et al. 2011; Laland et al. 2016; Odling-Smee et al. 2003). Whether this process is distinctive enough to earn niche construction status as a force of evolution is debatable, but nonetheless the concept of feedback between subsistence decisions and anthropogenic habitat modification is important for tracking historical change. This is true particularly in light of the fact that food production as consistent practice emerges from this feedback process, which can amplify the (initially small) improvements in resource yields and/or predictability that human activities stimulate.

Evolutionary ecology offers a unifying framework for developing functional explanations of subsistence choices in environmental context (Broughton et al. 2010). However, it is not designed to analyze the historical processes and mechanisms of change. For insight into these mechanisms and how they operate over time, I consider the role of social learning in the differential perpetuation of norms and behaviors. Social learning, and thus cultural transmission, is responsible for the spread of innovations. Whether a new crop or a new mode of production, an innovation will remain insignificant if it is not perpetuated by learning and imitation. Often cultural transmission supports functionally superior alternatives, but does so at varying rates depending on patterns of social interaction, population size, and operation of noneconomic decision criteria (Henrich 2001a, b; Henrich et al. 2005).

Types of Transitions in the Development of Food Production

The history of food production in North America across more than 10,000 years of development presents at least three types of transitions that are evident in the archaeological record. These include the following:

1. INITIAL DOMESTICATION OF LOCALLY AVAILABLE SPECIES. The domestication of seed crops in the midcontinental United States is reasonably well understood as an ecological process, but considerable disagreement exists about the economic utility of the earliest domesticates. A particular point of contention is whether initial domestication was related to declining foraging efficiency in the context of resource depression (depletion, especially when caused by human predation) or simply an opportunistic strategy intended to improve yields and reliability.

2. INITIAL ADOPTION OF DOMESTICATES AND AGRICULTURE FROM EXTERNAL SOURCES. In the Southwest, the earliest agriculture was based on an exotic crop plant from Mesoamerica. This situation mirrors that of populations that took up maize farming in parts of the Southeast where native seed crop cultivation was of minor economic importance. In the Southeast, maize seems to have been the first important domesticate. Its adoption therefore represented an initial transition to an agricultural economy rather than a shift to emphasize a particular crop. The initial adoption model also fits populations of the Eastern Woodlands living outside the natural range of floodplain weeds and other annual plants of the EAC. In these cases, the crop in question (and perhaps information about its use) was introduced via human agency. Under these circumstances, although agriculture may have presented itself as a complex of crops and associated knowledge rather than emerging from interactions with local plant populations, similar decisions had to be made about how much labor to invest in food production, when and where to grow crops, and what methods to use. These

economic decisions were contingent upon the availability of domesticates through exchange and other forms of social interaction.

Three events in this category are of particular interest: the adoption of maize in the Greater Southwest, the acquisition of Old World crops and livestock by Native groups, and the adoption of maize by midwestern EAC farmers. In none of these cases did novel domesticates immediately transform Native agricultural systems. In postcontact North America, groups buffered from direct contact with Europeans were selective, emphasizing crops that were low in cost, entailed little risk, and could be incorporated into existing cropping systems piecemeal rather than entailing a "whole-package" adoption of crops, tools, and techniques. Rapid acculturation to European farming methods was limited to planned settlements, such as Spanish missions. Autonomous Native communities were more likely to reject innovations such as livestock raising, plowing, and cereal crops such as wheat, although these had varying success depending on their utility in different settings. Maize also had varying degrees of success as a new crop in the East, persisting for centuries as a minor crop of limited distribution before becoming a staple.

3. INTENSIFICATION OF PRODUCTION. The rapid intensification of maize farming that took place in the East around AD 800 contrasts with the gradual increase in dependence on native seed crops that began after 1000 BC. Nonagricultural food production also represents a form of intensification. Management of perennials and aquatic resources in the Pacific Northwest, for example, involved significant labor investment and establishment of use rights, yet without apparent domesticates or maintenance of anthropogenic patches devoted to crop production. Intensification was not inevitable, as demonstrated by the incorporation of maize into subsistence economies as a fallback food or trade item with a minimal investment of labor.

Initial Domestication of Locally Available Species

The crops of the EAC and their contemporary free-living relatives are ecologically weedy, adapted to disturbed habitats where they have a competitive advantage. Although they have expanded their range to colonize anthropogenic habitats, three of the four earliest known domesticated plants of the region (marshelder, goosefoot, and *ovifera* squash) originally evolved to exploit the naturally disturbed habitats of floodplains (see Chapter 2). Bruce

Smith, in a series of publications, outlined the process by which floodplain weeds evolved into domesticates in the context of intensified human settlement (Smith 1987b, 1992b).

This scenario offers a plausible narrative for explaining how humans and plants entered into a coevolutionary relationship. However, it does not explain why humans began to save seed stock for planting. It is this practice that drives domestication by ensuring that harvested seeds provide the gene pool of subsequent generations. Without it, any sampling bias created by harvesting (such as overrepresentation of larger, non-dormant seeds) is not perpetuated, and directional selection for domestication-related traits never gets off the ground. So why plant? Smith envisions broadcast sowing, a nearly cost-free method of creating stands of annual plants conveniently located for future harvesting. The Natchez people of the lower Mississippi Valley practiced this type of casual cultivation in the eighteenth century, which would have had considerable utility because of the unpredictable shifts in the location of natural stands from year to year (Smith 1992d). Because the plants involved are so tolerant of a wide range of environmental conditions, they need not be tended throughout the growing season; broadcast sowing involved minimal labor with a large payoff, namely a dependable food source in a known location. Planting would also have allowed people to relocate stands in upland areas away from the floodplain, enhancing the reproductive isolation of the domesticate gene pool from free-living populations.

The floodplain weed theory does not take up the question of why, in a habitat with abundant relatively low-cost and high-yielding resources (such as impounded shallow-water fish, migratory waterfowl, and whitetail deer), humans would have chosen the small seeds of the EAC as a food source at all. Although small seeds have the potential to be highly productive under favorable conditions (Smith 1987a), preparation for consumption is often a very time-consuming operation depending on the method and degree of processing. Smith's yield estimates for goosefoot (1987a), although they are corrected to account for harvested material that does not provide nutrients (seed coat and plant parts usually labeled "chaff"), do not factor in processing costs. When included with planting and harvesting as an element of handling costs, processing can greatly reduce the net returns expected from harvesting small seeds. When compared to alternative food sources, such as large game and hickory nuts, small seeds are among the least profitable food items encountered by foragers worldwide (Gremillion 2004:Table 2). As long as these alternatives were abundant, time spent on small seed production would have been an economically inefficient option.

It makes sense to ask, then, how much labor was actually invested in processing of small seeds and how effective the technology was. Some direct evidence of seed processing comes from protected settings in dry caves and rockshelters, which have the potential to preserve paleofeces and any residues of on-site winnowing, threshing, or similar methods of removing inedible material. Paleofeces dating to the Late Archaic and Early Woodland periods from Kentucky often contain substantial quantities of seed coats ingested along with goosefoot seeds. Achenes of sunflower and marshelder were often consumed whole, although patterns of fragmentation are consistent with those created experimentally by pounding (Gremillion 2004). Light processing would have limited handling costs and made small seeds more profitable to exploit. Effective technology for cleaning extraneous material or releasing nutrients by cooking would also have improved return rates for small seeds. But bedrock mortars and basketry (which could have been used for winnowing) are associated for the most part with occupations postdating 1000 BC, too late to account for initial use of these food sources by human groups in the region.

Because of their high processing costs, EAC crops would appear near the bottom of a ranked list of resources based on profitability (net energy acquired per unit time spent). However, economic models of optimal resource selection demonstrate that incorporating such foods into the diet can actually improve foraging efficiency when more profitable items decline in abundance. The diet breadth or prey choice model, for example, quantifies the effects of scarcity on the time spent searching for food. Broadening the diet to include more items limits these search costs, which partly offsets the loss of efficiency imposed by turning to less profitable foods (Bird and O'Connell 2012; Winterhalder and Goland 1997; Winterhalder and Smith 2000). This trade-off may explain the broad-based diets that occur in the context of resource depression (depletion of prey populations due to human predation; Broughton et al. 2010).

Could resource depression caused by growing human populations have brought even these low-ranking foods into the optimal resource set? Although the logic behind demographically driven explanations is compelling, it is difficult to test them using archaeological data. According to

Smith and Yarnell (2009), there is no evidence for population "packing" in the river valleys where the first domesticates occur. They cite as evidence for this claim the generous spacing between prehistoric resource catchments, but the question remains as to why the magnetic pull of resource-rich river valley locations did not result in competition for resources. A challenge to the "stress-free" model of initial domestication comes from a recent analysis of radiocarbon dates and site frequencies for the midcontinental zone of early domestication. This study identified an increase in human populations between 4450 BC and 3050 BC, which was followed by a period of stasis from 3050 to 1850 BC. Weitzel and Codding (2016) interpret this stasis as demographic stabilization following initial domestication, which they argue was a response to resource stress. The data currently available, then, situate the earliest known domesticates (at ca. 3000 BC) near the end of a period of demographic growth. Whether resource competition is causally related to the utilization and domestication of small-seeded annual plants remains an open question. However, the demographic data indicate that this possibility deserves consideration.

Winterhalder and Goland (1997) discuss other solutions to declining foraging efficiency that may have altered the economic potential of the EAC. These include technological innovations that allow more efficient processing and habitat enrichment that increases the density of seed-producing resource patches (and thus their profitability) or boosts yields. Niche construction activities may take the form of inadvertent tweaks to local ecology that are by-products of other activities (trampling, clearing for construction, trash deposition) as envisioned by Smith (1987b). However, it seems unlikely that this recalibration of the cost-benefit ratio would have been sufficient to close the rather large gap in return rates between small seeds and arboreal nuts such as hickory (see Table 2.2).

This analysis of resource rankings assumes simultaneous availability; it does not account for seasonal fluctuations. By considering the "menu" of items by season, it is easy to see how small seeds might come to outrank other food resources periodically, if not on an ongoing basis. Such a shift in relative rankings is likely to have occurred outside of the prime harvest season in late summer and fall, when hickory nuts, acorns, walnuts, fleshy fruits, and animals in prime condition were available. In contrast, winter and spring were lean times for prehistoric foragers in the Eastern Woodlands. Game was available but not in prime condition, and plants largely dormant and still months away from flowering and fruiting. This situation of seasonal scarcity can be effectively addressed by storing surplus harvests for later use (Cowan 1985a; Smith and Cowan 2003). This factor, the possibility of deferred use, is not considered in the diet breadth model. However, relaxing the assumption that food is consumed immediately once acquired has important implications. One of these is that deferral entails some risk of loss, but at the same time serves to enrich the potential resource base for winter and spring. Stored seeds are more likely to offer relatively good return rates than newly harvested seeds because of the limited availability of alternate resources during winter. That they were used in this way is supported by archaeological evidence of caches and containers in sheltered locations (Fritz 1997; Smith 1985a). Contents of human paleofeces also indicate that crops were consumed out of season (Gremillion and Sobolik 1996).

Storage in the context of seasonal variation in food supply solves the puzzle of why small seeds were used at all in an environment richly endowed with superior alternatives. It also suggests that production of seed crops was at least in part an attempt to reduce intra-annual variance in food supply. They were not the only storable plant resources available, nor were they the most profitable; most tree nuts provide better return rates. However, hickories, oaks, and many other trees produce seeds by "masting," producing heavy crops at irregular intervals and failing to produce at all in some years (Gardner 1997; Kelly and Sork 2002). Masting is thought to be an adaptation to predation because superabundance satiates predators in good years, leaving some seeds to germinate. The consequence for human predators is that it is difficult or impossible to predict the most promising fall foraging locations. In contrast, seed crops are reliable annual producers and, if planted rather than naturally dispersed, they will be available at a known location. Seed caches dispersed strategically on the landscape in secure, sheltered sites would have helped to sustain small groups on short visits to upland camps (Cowan 1985a), including the ones associated with the floodplain sites that have yielded the earliest domesticates (Smith 2011a). That this function of seed crops is not more clearly indicated archaeologically may simply be an artifact of differential preservation.

Initial Adoption of Domesticates from External Sources

Maize in the Greater Southwest

In the Greater Southwest, desert foragers became farmers through adoption of a non-native plant by way of mechanisms that are still poorly understood. Both migration and down-the-line exchange are plausible routes of introduction, but archaeological evidence has not been very informative about their respective roles and importance. The best cases of possible migration of maize agriculturalists from the south occur in the southern part of the region, which makes sense given the Mesoamerican source of the plant. However, the primary mechanism for the spread of maize seems to have been the movement of goods and information between groups rather than the displacement of hunter-gatherers by farmers.

Functional explanations for the initial adoption of maize in the Southwest must confront the question of what maize had to offer foraging groups. Estimated return rates for early maize are higher than the majority of wild plant foods available in southern Arizona (Diehl 1997). In that case, we might expect it to replace at least some of the low-ranking wild foods. This is not the case in Early Agricultural period communities of southern Arizona, where use of low-ranked foods continued for at least two millennia following adoption of maize (Vierra and Ford 2007). Perhaps there was some constraint that limited investment in its production, or early maize was not as profitable as later varieties.

Rather than offering a boost to overall foraging efficiency or harvest yields, maize may have been valued for providing a predictably located supply of storable food that served to offset seasonal shortages (Diehl and Waters 2006). Wills (1995) suspects that risk reduction played a role in the adoption of maize in the Mogollon Highlands and Colorado Plateau. He argues that maize was used in the Mogollon Mountains during spring (a period of low availability of wild resources) to supplement hunting and cached in upland rockshelters to maintain control over productive piñon locations. Maize was valuable primarily because it improved hunting efficiency and allowed longer stays at distant locations. While long-distance hunting may not have been a viable strategy (Diehl 1997), caching of maize in rockshelters permitted forager-farmers to focus their efforts on highly productive (nonagricultural) resource patches. Perhaps maize offered a greater boost to efficiency in southern Arizona where growing conditions were better for maize and alternative plant foods provided low foraging returns (Wills 1995).

Adoption of maize was probably also facilitated by the preexisting practice of wild seed harvesting. Weedy plants such as goosefoot and amaranth were processed, consumed, and perhaps planted or tended during the Middle Archaic period (see Chapter 3). Seed harvesting and processing techniques were already in place that could be adapted for the new crop. Adams's experiments with stone grinding tools (1999) indicate that the same tools served well for both seeds and early maize, although the introduction of flour corn inspired some design changes.

Initial Adoption of Maize in the East

Reasons for the initial adoption of maize in the East are difficult to assess in part because the earliest occurrences are few and scattered geographically (see Chapter 4). With the exception of the phytolith residues from New York (at ca. 300 BC; Hart et al. 2007), the earliest maize in the region comes from the area within which the Hopewellian exchange system was active. It is reasonable to suppose that maize traveled along the same trade routes that brought obsidian from Yellowstone, mica from North Carolina, and Great Lakes copper to central Ohio (DeBoer 2004). Perhaps it was originally an exotic item with prestige value that was introduced as a commodity. These persistent questions make it difficult to say with confidence what the status of maize was in subsistence economies. Perhaps it was introduced and failed to spread, which would explain its spotty archaeological occurrence before AD 800. Reasons for its adoption are not clear.

Adoption of Old World Crops and Livestock

The massive influx of previously unknown biota into North America after 1492 presents a very different situation because of the cultural gap between recipient and donor populations. Native peoples struggled with altered networks of social interaction, novel power structures that placed them at a disadvantage, and devastating "virgin soil" epidemics (see Chapter 6). In frontier zones the dynamics of culture contact and the magnitude of change make it difficult to untangle the causal threads that resulted in some of these novelties being adopted and others being ignored. Where indirect trade with European sources was the prevalent form of contact, the availability of potential new crops is largely unknown, so the apparent conservatism of these groups may be attributable instead to a lack of knowledge. Where this knowledge was available, patterns of adoption of agriculture, new crops, and livestock share some striking similarities in very different cultural, economic, and environmental contexts. One shared theme is that in the absence of coercive acculturation, such as took place in mission settings, the response to newly introduced crops was typically selective. The nature of that selectivity provides some important clues as to the decision criteria at work.

Newsom and Gahr (2011) list a set of features shared by the introduced crops that were accepted readily by Native groups. Referring to Gremillion (1993a) and Lopinot (1986), they identify these as 1) ecological weediness; 2) suitability and compatibility with existing subsistence activities; 3) similarity to native crops; 4) a relatively high ratio of benefits to costs; and 5) relatively low risk of adverse outcomes. For a crop plant to be categorized as "weedy," it must have some ability to survive apart from humans; the term also implies broad environmental tolerance and a tendency to colonize disturbed habitats. Such traits allow crop plants to survive adverse conditions such as neglect and make them more likely to naturalize. Many of the other traits in the list are related, directly or indirectly, to economic considerations. Compatibility speaks to opportunity costs; activities that can be freely pursued without taking time away from other important tasks are more likely to be adopted. Similarity to native crops seems to be a good predictor of suitability to the local environmental conditions, but it also suggests an existing store of cultural knowledge that might be applied to its cultivation. Limiting investment in innovation to non-staple resources also serves to restrict the impact of adverse outcomes.

These generalizations emphasize the time and labor costs associated with producing a new crop and how they compare with the yields obtained. However, a closer look at this list of shared traits implicates the costs of learning as well. As a species, we have evolved mechanisms of social learning that are often less costly than individual learning through experimentation (Boyd and Richerson 1985). However, social learning relies on contacts between people that facilitate observation, imitation, and information exchange. In the absence of such contacts, the costs of individual learning about the new resource can be high enough to discourage adoption. In such cases, the effective use of the new item or technique relies on access to bodies of cultural knowledge that have accumulated over generations. Socially acquired technological information is much less crucial for transfer of novelties that can easily be learned about through experimentation, either because similar resources are already in use in the recipient population, or because the resource has inherent properties that make it easy to produce, prepare, and consume without specialized knowledge.

It is probably not coincidental that so many of the Old World crops that were taken up by Native communities came to them by way of mission settlements, which provided instruction in cultivation and preparation of unfamiliar crops as well as the requisite seeds, tools, and even draft animals (Newsom and Gahr 2011). Learning and observation in this context supplied knowledge that could be shared with others, perhaps making adoption more likely than it would have been otherwise. For other crops, such social learning was not required because traditional knowledge sufficed; for example, the watermelon was similar enough to native cucurbits that any relevant cultivation techniques could be transferred (Blake 1981). The same may be true of introduced legumes in the East and the Southwest, and of the Andean potato and turnips on the Northwest Coast, which resembled native geophytes. In the case of potatoes, intermarriage between Native women and fur traders provided an additional stimulus to adoption (Newsom and Gahr 2011).

In the case of domestic animals, particularly livestock, there were no analogous traditions that could have provided relevant cultural knowledge. Adult consumption of milk products was unknown, and the vast majority (some 90%) of Native Americans were lactose intolerant, unable to effectively digest most dairy products (Pavao-Zuckerman and Reitz 2011:580). Agricultural systems that might accommodate new and relatively profitable crops could not incorporate livestock production without considerable reorganization of labor and reallocation of resources. Livestock need year-round attention, limiting the ability to travel freely, and need to be protected from predation and other losses. Free-ranging stock can destroy crops and damage vegetation and soils through trampling, coming into direct conflict with farming (Pavao-Zuckerman and Reitz 2011).

Intensification of Production

Intensification refers to increased production per unit land (this definition has been contested, but has the advantages of clarity and simplicity). Typically this process involves irrigation, enrichment of the plants' substrate (for example, by using fertilizer), reduction of competition by weeding, and other inputs that increase yields per unit area of land. These criteria apply to forms of food production other than agriculture, such as burning and pruning perennials to stimulate growth. Because it requires greater input of labor, intensification does not necessarily improve efficiency and in many cases lowers it (Ames 2005; Boone 2002). There is a nonlinear relationship between labor input and yields, producing a diminishing rate of increase over time. Some researchers maintain that a decline in efficiency is inevitable with intensification and therefore it is not likely to occur unless forced by a need for greater absolute quantities of food, as might be the case for a growing population (Thurston and Fisher 2007a, 2007b). Other "push" factors include the imposition of tribute or taxation, social obligations, and prestige competition.

I begin with the assumption that intensification is not an inevitable process, but one that becomes an attractive option because it offers some anticipated benefit. That benefit can take the form of more efficient energy capture (unlikely because of diminishing returns on investment), reduced variance (which limits risk), or the production of surplus to invest in strengthening social bonds that support cooperation or accumulation of wealth and power.

The discussion of initial domestication and adoption of crops above asks how inclusion of these novelties in a group's subsistence repertoire is likely to affect efficiency and risk. Intensification involves a further commitment to increase labor investment in crop production. This transition may occur immediately, after a long delay, or may never take place at all, as when low-level food production proves to be a stable adaptation (Smith 2001). The question of what factors influence these varied outcomes has preoccupied archaeologists for decades.

Under what conditions are people likely to increase their investment in an activity that boosts the productivity of land even as it undermines economic efficiency? The most obvious candidate cause is that the anticipated food supply falls short of perceived needs. With energy limited in this way, it makes sense to invest available time in food acquisition; when time is limited but not energy, people tend to invest it instead in other important activities (Bettinger 2001). One pathway to the energy-limited situation is by way of resource depression and competition between human populations (Bettinger 2001). When competition places constraints on mobility as a solution to food shortage, as often happens in the context of local or regional population growth, intensification may offer an alternative way to meet needs.

Shifting utility of time and energy also comes into play when crops are produced for storage. Storage can offset annual shortfalls caused by seasonal variation. It also has the potential to limit vulnerability to longer periods of deprivation resulting from unpredictable interannual variation in environmental conditions. However, future value of stored goods is somewhat discounted for a variety of reasons, most notably the increasing probability of loss with continued delay in consumption (Tucker 2006). This perspective highlights the ability to fine-tune labor inputs as conditions change. Processing can be deferred until winter, when energy is most limited and plant foods scarce.

Native Seed Crops in the East

The intensification of seed crop production in eastern North America after 1000 BC seems to have been a gradual and regionally variable phenomenon. The timing of this transition varies, with substantial increases in the deposition of domesticated and cultivated plants in archaeological deposits appearing earliest in favorable preservation contexts (for example, some of the rockshelters of the Ozarks and Cumberland Plateau). High counts of seed crops peak in the Early Woodland of eastern and central Tennessee and west-central Kentucky, the Middle Woodland of the lower Illinois Valley, and the Late Woodland of the American Bottom, central Ohio, and central Arkansas (Johannessen 1984, 1993; Smith and Cowan 2003; Yarnell and Black 1985). In many parts of the East, however, native seed crops never achieved much importance and appear in the archaeological record only in relatively small quantities. This pattern of limited use is evident along much of the Atlantic and Gulf coastal plains, including the lower Mississippi Valley.

Variation in the economic role of native seed crops suggests the operation of causal factors that were historically or environmentally contingent at the local and subregional levels. The EAC crops can produce seed under a broad range of environmental conditions even with limited care, but their value was related to the benefits of storage for seasonal use. If EAC crops were highly ranked only in the winter and spring seasons of scarcity, we would expect a correlation between harvest yields and length of the growing season. Mapping of relevant archaeobotanical data available in 2002 does indicate that the region of greatest reliance on native seed crops falls largely within an area that today has between 60 and 100 days in which the temperature drops below freezing (Gremillion 2002a). This pattern is fairly robust for data available in 2002, remaining unaffected by adjustments for sample size, differential preservation, and other possible confounding variables, although it may change as new archaeobotanical data from the interior Southeast comes to light (see discussion in Chapter 3). Although the study did not cover the Northeast, based on current evidence, the EAC played a minimal role at best north and east of the upper Ohio River valley (Crawford 2011). Reasons for this pattern are not clear, but they may reflect limited interaction between the Midwestern source area of the earliest domesticates (which is drained by the Mississippi and its tributaries) and regions whose rivers drain into the Atlantic (including much of the Northeast and Southeast).

Production for storage rather than immediate use has implications for the value of the harvest. When future need is certain, the best possible rate of return is achieved by storing only the highest ranked foods (the front-loaded strategy; Bettinger 2009). Many of the seed crops cultivated on the Cumberland Plateau are heavily back-loaded (Bettinger 2009) because they are inexpensive to harvest but costly to process. For this reason they would have been useful as an inexpensive form of insurance when the probability of an actual shortfall was low (Cane 1989; Gremillion 2004). Doing so is an effective way to provide insurance against hunger without paying expensive processing costs up front; these need only be incurred if the cache is actually used. An added benefit is that deferral of arduous winnowing, parching, and grinding to the winter reduces opportunity costs incurred during the busy harvest season of the fall months (Gremillion 2004). Time spent processing food can be costly when other tasks are more urgent; putting off the arduous work of winnowing, grinding, and shelling makes economic sense if it can be done when such conflicts do not exist.

Winterhalder and Goland's (1997) application of the diet breadth model illustrates how the resource depression pathway might result in agricultural intensification. It shows how both population growth and resource depression can result from increasing reliance on domesticates, particularly if they are low-ranked (see also Winterhalder and Goland 1993). It is easy to envision a positive feedback loop in which the human population comes perilously close to the limits imposed by potential productivity of the environment. Under these conditions, intensifying production of domesticates is one way, perhaps the only way, to ensure an adequate supply of food. Winterhalder and Goland (1997:157) doubt that EAC crops were ever abundant enough to fuel population growth (maize is a different story; see below). However, several lines of evidence point to significant population increase during the period from 1400 to 500 cal BC in the midcontinental region where EAC crops were first domesticated (Weitzel and Codding 2016). This timing coincides with archaeobotanical evidence of intensification after 1000–500 BC, although the chronological and causal relationships between the two events remain unclear.

Maize Agriculture in the Southwest

Following its initial introduction as early as circa 2000 BC, maize remains occur in varied but generally modest quantities for a millennium or more before there is any indication of its increasing economic and dietary importance. Although there is disagreement about whether maize initially served as a staple or a minor supplement (and both may be true), there is a clear broad-scale trend to greater use over time. However, as is the case with EAC crops, the subsistence role of maize varies considerably within the region.

Early Agricultural period maize had smaller yields than later varieties, based on morphological attributes of archaeological cobs that are correlated with productivity. Potential maize yields increased continually throughout prehistory in southern Arizona, a trend that started as early as 500 BC (Diehl 2005). If maize yields were indeed relatively low at first, there would have been little impact on fertility or birthrates, but anything that increased those yields had the potential to initiate population growth. The resulting decline in overall efficiency coupled with constraints on mobility inspired the adoption of technological innovations—baskets for collecting and winnowing, grinding implements, effective preservation methods, and better cookware. Storage technology is implicated in the case of southeastern Arizona, where Diehl and Waters (2006) propose that it was only with the introduction of ceramic storage jars at AD 200, which cut losses of stored grain, that more intensive cultivation began to pay off economically. The introduction of flour corn varieties also boosted profitability by making processing less time-consuming, although it does not necessarily offer higher yields (Adams 2015). The floury variety *Mais de Ocho* was present at Tularosa Cave in southern New Mexico as early as AD 400 (Adams 2015:20). Its introduction coincided with some changes in the form of grinding equipment that further increased efficiency of maize use (Adams 1999; Diehl 1996). Increasing birth rates in the Southwest between 1000 BC and AD 500 may have created some pressure on foraged resources and driven efforts to improve processing efficiency (Kohler and Reese 2014).

These technological improvements increased the return rates obtained from maize, making it a first-line resource. Specialization on maize farming now offered significant economic rewards. Maize and a limited set of foraged resources became central to subsistence economies, along with opportunistic use of a wide range of less profitable fallback foods. This strategy resembles the Hohokam pattern. Hohokam farmers of the Tucson Basin (AD 500–1500) were more sedentary than their Cienega phase (Early Agricultural period, 800–200 BC) predecessors and had less diverse diets, having opted to concentrate on maize and other domesticates at the expense of wild or weedy annuals (Diehl 1997). However, wild and weedy plants and animal foods remained important in the diets of most southwestern groups. Foraged resources probably played a key role in an environmental setting in which the chief limiting factors in maize cultivation—moisture and length of the growing season—were highly variable across time and space. This variation may be masked by assemblages that are actually aggregates of many annual cycles and may conceal variation in time, such as intermittent use of famine foods, or between groups with different foraging ranges (Bettinger 2001).

Most discussions of maize intensification focus on relatively long-term (century- or even millennium-scale) trends to increasing production. In some cases, however, variability in maize use between contemporaneous sites is more evident than any temporal trend. The Fremont farmers of the southern Great Basin, adjacent to the Southwest proper, present such a case. Desert foragers who adopted maize, the Fremont seem to have maintained a flexible strategy that allowed them to vary the investment of labor in farming depending on the local availability of nonagricultural resources (Barlow 2002, 2006). Barlow compares marginal return rates for four different agricultural strategies (plant and harvest, slash-and-burn cultivation, "typical" agriculture, and intensive agriculture) to average returns for foraged wild foods in the Fremont area. This comparison shows that maize farming can provide yields equal or superior to many foraged resources, but only if labor investment is limited to a simple "plant and harvest" strategy. Greater investment is unlikely to pay off unless most other resources are depleted and foraging efficiency is low. This finding is important because it shows that simply spending more time on producing maize does not necessarily reap impressive benefits, due to diminishing returns. Claims that high-yielding maize varieties rewarded increased production would receive some support if it could be demonstrated that the greater costs of field preparation, tending crops, and post-harvest processing would not have outweighed any increase in efficiency.

Intensification of Maize in the East

The millennium-long gap between the acquisition of maize and its intensification has long puzzled archaeologists (see discussion in Chapter 4). Assuming that low-level use continued despite the sparse archaeological record of maize before AD 750, what is responsible for the rapid intensification of maize production between AD 750 and 1150? The resource depression-population growth pathway is one candidate. The timing of the rapid intensification of maize follows a period of population growth that began at 1400 BC and ended at AD 550 (Weitzel and Codding 2016). Assuming that this population trend has been accurately identified, maize intensification seems to have commenced during a period of relative stasis or decline in the human population of the midcontinent, rather than in the midst of a regional increase (during which time maize was perhaps known but little-used).

An alternative pathway to intensification involves changes that made maize cultivation more economically viable by removing constraints on yields and/or efficiency of use. One such constraint might be the original adaptation of maize to temperate-zone photoperiodism. The modern cultivar Northern Flint is adapted to the long summer days at higher latitudes and will trigger flowering and seed production early enough to ensure maturation of the cobs (Bonzani and Oyuela-Caycedo 2006). However, Northern Flint is not unique in this regard. Fritz (1994) argues that reduction of photoperiod sensitivity necessary for flowering and fruiting occurred before maize moved into the East, based on its expansion into southern Colorado. Similarly, adaptation to shorter growing seasons would have been necessary at the higher altitudes it encountered during early dispersal into the Southwest. Therefore, it does not seem likely that diffusion of maize to the East was significantly delayed by temperate-zone photoperiodism or growing season length.

There is some evidence that the introduction of new varieties of maize with novel traits influenced patterns of consumption and production in

North America. The appearance of *Mais de Ocho* in the Southwest coincides with new types of grinding tools effective for creating flour (Adams 1999). In the East, another eight-row variety, Northern Flint, was well suited to making hominy and thereby improving the nutritional quality of maize (see below). The distinctive characteristics of Northern Flint, such as low row number and wide kernels and cupules (structures on the cob that hold the kernels), are represented in the type labeled Eastern Eight-Row in archaeobotanical collections (Wagner 1986). Eight-row maize is prevalent in the Northeast but not in the Midwest, where varieties with high row numbers dominate archaeobotanical collections (Wagner 1986). Thus, it is unlikely that the appearance of Eastern Eight-Row/Northern Flint drove the earliest efforts at intensification in the Midwest.

Technological innovations also have the potential to increase the utility of maize as a food. One such innovation is nixtamalization, the alkali treatment that improves the nutritional quality of the kernels. The product of nixtamalization is hominy, which consists of softened maize kernels that are relatively easy to process. Perhaps more significantly, this treatment releases the essential amino acids tryptophan and lysine, which would otherwise not be digested. Without nixtamalization, heavy dependency on maize can result in pellagra, a consequence of niacin deficiency (Briggs 2016). However, it is difficult to pinpoint the advent of this technology in the East. Although certain morphological traits distinguish kernels that have been processed into hominy and then subsequently carbonized (Dezendorf 2013; Lopinot 1994), the same treatment also greatly increases the chances of preservation. Thus, any processing of maize *without* alkali treatment is likely to underrepresent kernels, complicating the interpretation of its dietary role during its early history in the East.

Other technological advances in maize processing may have appeared earlier; for example, there is a correlation between pottery wall thickness and indicators of maize consumption in the Northeast that begins as early as AD 200. Thinner walls conduct heat more effectively than thick ones and better resist thermal stress, traits important for the long periods of direct-heat cooking associated with maize (Hart 2012). Similarly, the shell-tempered Moundville Globular Jar vessel form, well suited to the long periods of boiling, became the dominant form after AD 1120. The timing of this innovation coincides with the shift to greater reliance on maize in the Moundville area of west-central Alabama (Briggs 2016).

In the Deep South, east of the Appalachians and along the Gulf and Atlantic coastal plains, the lag between maize introduction and intensification is much shorter than it is in the midcontinent. Maize was present by AD 900 in west-central Alabama in small quantities, and within 200 years it had become a subsistence staple (Scarry 1993). Perhaps maize appeared at a time when foraged resources were scarce, and especially if relatively high in rank, rapid intensification would be expected. The process would have been accelerated by any deterioration of local habitats. Anderson and colleagues (1995) explored this possibility by documenting rainfall variability in the Savannah River valley between AD 1100 and 1500. They suggest that multi-year droughts created agricultural shortfalls, but they also would have impacted foraged resources such as acorns, hickory nuts, small seeds, and fleshy fruits. These frequent fluctuations seem to call for a flexible strategy in the manner of the Fremont farmers described by Barlow (2006).

Most of the explanations proposed to account for intensification emphasize the economic costs and payoffs for individuals or households. However, sociopolitical dynamics also have the potential to motivate change in subsistence regimes. Maize intensification was associated with, and apparently coevolved with, the development of the transegalitarian societies labeled Mississippian by archaeologists. Whether intensification preceded the development of social ranking and suprahousehold political organization is not yet known, and the resolution of the archaeological record may be insufficient to answer the question. However, changes in social relations that allow elites to control food stores would have altered access to resources much in the same way that resource depression does, by reducing the availability of preferred foods and imposing high costs on their inclusion in the diet. Competition between polities can also drive intensification.

Intensification without Agriculture

Chapter 5 discussed several examples of food production without domesticates or agriculture as they are usually defined. In the Pacific Northwest, archaeologists are still seeking data needed to discover the time depth of practices that have been documented ethnohistorically or ethnographically. Collaborative research with Native communities indicates that resource management focused on perennials and mass animal resources such as clams rather than annual plants (Lepofsky et al. 2015; Smith 2005b). The plant parts that have been recovered do not show clear signs of domestication; this is not surprising since these tend to develop slowly in perennial species, which may not reproduce every year. Lack of obvious morphological indicators of domestication leads researchers to underestimate the impact of selection in anthropogenically modified habitats (Smith 2005b; Turner and Peacock 2005). The most important prehistoric evidence in support of food production here is in the form of substrate modification (e.g., preparation of clam beds and root gardens) and microcharcoal records of burning. Camas processing is known to have been practiced for millennia on the Columbia Plateau, providing evidence of intentional burning to maintain the camas, which would otherwise be replaced by other vegetation (Lepofsky et al. 2005). The timing of intensification and its causes are not well understood at present, but ongoing archaeological research may yet produce some answers.

In the Great Basin, food production took the form of low-investment sowing of seeds in floodplains, controlled burning, and pruning of perennial plants. Intensification of the small seed complex emphasized technological innovations rather than modifications arising as a result of domestication. Failure to move from small seed intensification to small seed domestication may be attributable to the high mobility of human groups in the Great Basin, which inhibited the establishment of the kind of persistently disturbed anthropogenic habitats that drove initial domestication in the East. Maize was a late introduction from external sources, and one that augmented but did not replace foraged foods. In fact, as long as the latter were abundant, intensive agriculture would have remained a relatively expensive and inefficient option.

Conclusion

North America is a vast continent, environmentally diverse and colonized by humans late in the Holocene. Compared to other independent centers of plant domestication on a global scale, agriculture was relatively late. Smith (2011a) argues that the delay occurred because mid-Holocene climate change was the trigger that initiated coevolution with native crops in the East, while others see a causal link with population growth (Weitzel and Codding 2016). Whatever stimulated the initial incorporation of small seeds into human diets, intentional planting and the practice of storage buffered fluctuations in food supply, created predictable resource patches, and allowed these plants to travel with people on their seasonal rounds. Husbandry of small seed-producing annual plants did not always result in domestication or fully agricultural economies, but nonetheless served to increase the predictability and productivity of plant resources. Similarly, in many areas perennial plants, which are not suited to the annual plant-and-harvest cycle of seed crop cultivation, were tended instead in their natural (but anthropogenically modified) habitats.

Maize was a transformative plant that eventually underwrote the production of surplus that could be stockpiled for future use, trade, or prestige competition. However, it remained a minor crop for centuries in most locations where it was grown. Reasons for its intensification continue to be debated. Diminishing returns on labor investment may have discouraged intensification initially, but it is not clear what changed to remove this obstacle. Candidates include technological innovations such as alkali treatment and better cooking pots, more productive maize varieties, and competition between emerging polities that privileged increased production over efficiency.

The Native peoples of North America together created a diverse set of practices and traditions devoted to the plants they harvested, tended, and cherished. Like people everywhere, they accumulated bodies of knowledge that could be put to work not only acquiring naturally occurring wild foods but also enhancing them. This practice, which we call food production, took many forms across the continent and across time. In some cases food production became a type of agriculture, but in many cases it did not. There was room for this diversity in a continent not yet overrun by vast fields of genetically identical maize and wheat. Movements are afoot to bring diversity and sustainability back to modern agriculture, and perhaps will lead to a greater appreciation of ancient agricultural traditions like those of Native America.

Appendix A. Scientific names of plant and animal taxa mentioned in the text.

Common name	Scientific name
African rice	*Oryza glaberrima*
Agave	*Agave* spp.
Amaranth	*Amaranthus* spp.
Andean potato	*Solanum tuberosum*
Apple, crab apple	*Malus pumila*
Apricot	*Prunus armeniaca*
Asian rice	*Oryza sativa*
Barley	*Hordeum vulgare*
Basswood	*Tilia* spp.
Beaver	*Castor canadensis*
Beech	*Fagus* spp.
Bighorn sheep	*Ovis canadensis*
Black bear	*Ursus americanus*
Black walnut	*Juglans nigra*
Blazingstar	*Mentzelia* spp.
Blueberry	*Vaccinium* spp.
Bluedicks	*Dichelostemma capitatum*
Bulrush	*Juncus* spp.
Butternut (White walnut)	*Juglans cinerea*
Butternut squash	*Cucurbita moschata*
Cactus	Cactaceae (Cactus family)
Camas	*Camassia* spp.
Cantaloupe	*Cucumis melo*
Cattle	*Bos taurus*
Cherry	*Prunus* spp.

Common name	Scientific name
Chicken	*Gallus gallus domesticus*
Chickpea (Garbanzo bean)	*Cicer arietinum*
Chilies	*Capsicum* spp.
Chocolate lily	*Fritillaria biflora*
Cholla	*Opuntia* spp.
Clams	Class Bivalvia
Common bean	*Phaseolus vulgaris*
Common dandelion	*Taraxacum officinale*
Cosmopolitan bulrush	*Bolboschoenus maritimus*
Cotton	*Gossypium* spp.
Cottontail	*Sylvilagus* spp.
Cottonwood	*Populus* spp.
Cowpea	*Vigna unguiculata*
Creosote bush	*Larrea tridentata*
Cucumber	*Cucumis sativus*
Curly dock	*Rumex* spp.
Cushaw squash	*Cucurbita argyrosperma*
Devil's claw	*Proboscidea parviflora*
Douglas fir	*Pseudotsuga menziesii*
Dropseed	*Sporobolus* spp.
Elk	*Cervus elephus*
Erect knotweed	*Polygonum erectum*
Fir	*Abies* spp.
Flat sedge	*Cyperus esculentus*
Flax	*Linum usitatissimum*
Fritillary	*Fritillaria* spp.
Garlic	*Allium sativum*
Garry oak	*Quercus garryana*
Giant ragweed	*Ambrosia trifida*
Goat	*Capra hircus*
Goosefoot (Lambsquarters)	*Chenopodium* spp.
Goosegrass	*Eleusine indica*
Grizzly bear	*Ursus arctos*
Groundnut	*Apios americana*
Guinea corn	*Sorghum* spp.
Hazelnut	*Corylus* spp.

Common name	Scientific name
Hemp	*Cannabis sativa*
Hickory	*Carya* spp.
Horse	*Equus caballus*
Indigo	*Indigofera* spp.
Jackbean	*Canavalia ensiformis*
Joshua tree	*Yucca brevifolia*
Lentil	*Lens culinaris*
Lettuce	*Lactuca sativa*
Lima bean	*Phaseolus lunatus*
Little barley	*Hordeum pusillum*
Lupine	*Lupinus* spp.
Maize	*Zea mays* ssp. *mays*
Mangrove	*Avicennia* spp., *Rhizophora* spp.
Maple	*Acer* spp.
Marshelder (Sumpweed)	*Iva annua*
Maygrass	*Phalaris caroliniana*
Mesquite	*Prosopis* spp.
Mexican crucillo	*Condalia warnockii*
Moose	*Alces alces*
Mountain goat	*Oreamnos americanus*
Mule deer	*Odocoileus hemionus*
Nettles	*Urtica* spp.
Northern rice-root	*Fritillaria camschatcensis*
Oak	*Quercus* spp.
Oats	*Avena sativa*
Olive	*Olea europaea*
Onion	*Allium cepa*
Opossum	*Opossum opossum*
Orange	*Citrus* spp.
Paloverde	*Parkinsonia* spp.
Panic grass	*Panicum* spp.
Pea (English pea, Garden pea)	*Pisum sativum*
Peach	*Prunus persica*
Pig	*Sus scrofa*
Pine	*Pinus* spp.
Piñon	*Pinus edulis*

Common name	Scientific name
Pitseed goosefoot	*Chenopodium berlandieri*
Plantain	*Plantago* spp.
Plum	*Prunus* spp.
Pokeweed	*Phytolacca americana*
Prairie dog	*Cynomys* spp.
Prickly mallow	*Sida spinosa*
Prickly pear	*Opuntia* spp.
Pronghorn antelope	*Antilocapra americana*
Raccoon	*Procyon lotor*
Radish	*Raphanus sativus*
Ricegrass	*Achnatherum hymenoides*
Rye	*Secale cereale*
Sagebrush	*Artemisia* spp.
Saltbush	*Atriplex* spp.
Seepweed	*Suaeda* spp.
Sheep	*Ovis aries*
Shepherd's purse	*Capsella bursa-pastoris*
Sorghum	*Sorghum bicolor*
Spikerush	*Eleocharis* spp.
Springbank clover	*Trifolium wormskioldii*
Spruce	*Picea* spp.
Squash	*Cucurbita* spp.
Sumac	*Rhus* spp.
Tamarack	*Larix laricina*
Teosinte	*Zea mays* ssp. *mexicana*
Tepary bean	*Phaseolus acutifolius* var. *latifolius*
Threesquare	*Schoenoplectus pungens*
Tobacco	*Nicotiana* spp.
Tobacco, domestic	*Nicotiana tabacum, N. rustica*
Tomato	*Solanum lycopersicum*
Turkey	*Meleagris gallopavo*
Turnip	*Brassica rapa*, Rapifera group
Turtles	Order Testudines
Wapato	*Sagittaria latifolia*
Watermelon	*Citrullus lanatus*
Wheat	*Triticum aestivum*

Common name	Scientific name
Whitestem blazingstar	*Mentzelia albicaulis*
Whitetail deer	*Odocoileus virginianus*
Wild onion	*Allium* spp.
Willow	*Salix* spp.
Wine grape	*Vitis vinifera*
Yellow glacier lily	*Erythronium grandiflorum*

References Cited

Adair, James
  1775 *The History of the American Indians.* E. and C. Dilly, London.
Adair, Mary
  2003 Great Plains Paleoethnobotany. In *People and Plants in Ancient Eastern North America*, edited by Paul E. Minnis, pp. 258–346. Smithsonian Books, Washington, DC.
Adair, Mary, and Richard R. Drass
  2011 Patterns of Plant Use in the Prehistoric Central and Southern Plains. In *The Subsistence Economies of Indigenous North American Societies: A Handbook*, edited by Bruce D. Smith, pp. 307–352. Smithsonian Institution Scholarly Press, Washington, DC.
Adams, Jenny L.
  1999 Refocusing the Role of Food-Grinding Tools as Correlates for Subsistence Strategies in the U.S. Southwest. *American Antiquity* 64:475–498.
Adams, Karen R.
  2008 Anthropogenic Ecology in the American Southwest: The Plant Perspective. In *Movement, Connectivity, and Landscape Change in the Ancient Southwest: The 20th Anniversary Southwest Symposium*, edited by Margaret C. Nelson and Colleen A. Strawhacker, pp. 119–140. University of Colorado Press, Boulder.
  2014 Little Barley Grass (*Hordeum pusillum* Nutt.): A Prehispanic New World Domesticate Lost to History. In *New Lives for Ancient and Extinct Crops*, edited by Paul E. Minnis, pp. 139–179. University of Arizona Press, Tucson.
  2015 The Archaeology and Agronomy of Ancient Maize (*Zea mays* L.). In *Traditional Arid Lands Agriculture: Understanding the Past for the Future*, edited by Scott E. Ingram and Robert C. Hunt, pp. 15–53. University of Arizona Press, Tucson.
Adams, Karen R., and Suzanne K. Fish
  2011 Subsistence through Time in the Greater Southwest. In *The Subsistence Economies of Indigenous North American Societies: A Handbook*, edited by Bruce D. Smith, pp. 147–183. Smithsonian Institution Scholarly Press, Washington, DC.
Ambrose, Stanley H.
  1987 Chemical and Isotopic Techniques of Diet Reconstruction in Eastern North America. In *Emergent Horticultural Economies of the Eastern Woodlands*, edited by William F. Keegan, pp. 87–107. Occasional Papers No. 7. Center for Archaeological Investigations, Southern Illinois University, Carbondale.

REFERENCES CITED

Ambrose, Stanley H., Jane Buikstra, and H. W. Krueger
　2003　Status and Gender Differences in Diet at Mound 72, Cahokia, Revealed by Isotopic Analysis of Bone. *Journal of Anthropological Archaeology* 22:217–226.

Ames, Kenneth M.
　2005　Intensification of Food Production on the Northwest Coast and Elsewhere. In *Keeping It Living: Traditions of Plant Use and Cultivation on the Northwest Coast of North America*, edited by Douglas Deur and Nancy J. Turner, pp. 67–100. University of Washington Press, Seattle.

Anderies, John M., Ben A. Nelson, and Ann P. Kinzig
　2008　Analyzing the Impact of Agave Cultivation on Famine Risk in Arid Pre-Hispanic Northern Mexico. *Human Ecology* 36:409–422.

Anderson, David G., David W. Stahle, and Malcolm K. Cleaveland
　1995　Paleoclimate and the Potential Food Reserves of Mississippian Societies: A Case Study from the Savannah River Valley. *American Antiquity* 60:258–286.

Anderson, Edgar
　1952　*Plants, Man, and Life*. University of California Press, Berkeley.
　1956　Man as a Maker of New Plants and New Plant Communities. In *Man's Role in Changing the Face of the Earth*, edited by William L. Thomas, pp. 763–777. University of Chicago Press, Chicago.

Anderson, M. Kat
　2005　*Tending the Wild: Native American Knowledge and the Management of California's Natural Resources*. University of California Press, Berkeley.

Anyon, Roger, Darrell Creel, Patricia A. Gilman, Steven A. LeBlanc, Myles R. Miller, Stephen E. Nash, Margaret C. Nelson, Kathryn J. Putsavage, Barbara J. Roth, Karen Gust Schollmeyer, Jakob W. Sedig, and Christopher A. Turnbow
　2017　Re-Evaluating the Mimbres Region Prehispanic Chronometric Record. *Kiva* 83:316–343.

Anyon, Roger, Patricia A. Gilman, and Steven A. LeBlanc
　1981　A Reevaluation of the Mogollon-Mimbres Archaeological Sequence. *Kiva* 46:209–225.

Asch, David L., and Nancy B. Asch
　1985　Prehistoric Plant Cultivation in West-Central Illinois. In *Prehistoric Food Production in North America*, edited by Richard I. Ford, pp. 149–203. Anthropological Papers No. 75. Museum of Anthropology, University of Michigan, Ann Arbor.

Asch, Nancy B., and David L. Asch
　1978　The Economic Potential of *Iva annua* and Its Prehistoric Importance in the Lower Illinois Valley. In *The Nature and Status of Ethnobotany*, edited by Richard I. Ford, pp. 301–341. Anthropological Papers No. 67. Museum of Anthropology, University of Michigan, Ann Arbor.

Axtell, James
　1988　At the Water's Edge: Trading in the Sixteenth Century. In *After Columbus: Essays in the Ethnohistory of Colonial North America*, edited by James Axtell, pp. 144–181. Oxford University Press, New York.

Baden, William M., and Christopher S. Beekman
  2001 Culture and Agriculture: A Comment on Sissel Schroeder, Maize Productivity in the Eastern Woodlands and Great Plains of North America. *American Antiquity* 66:505–516.

Baker, Herbert G.
  1974 The Evolution of Weeds. *Annual Review of Ecology and Systematics* 5:1–24.

Barlow, K. Renee
  2002 Predicting Maize Agriculture among the Fremont: An Economic Comparison of Farming and Foraging in the American Southwest. *American Antiquity* 67:65–88.
  2006 A Formal Model for Predicting Agriculture among the Fremont. In *Behavioral Ecology and the Transition to Agriculture*, edited by Douglas J. Kennett and Bruce Winterhalder, pp. 87–102. University of California Press, Berkeley.

Barrier, Casey R.
  2011 Storage and Relative Surplus at the Mississippian Site of Moundville. *Journal of Anthropological Archaeology* 30:206–219.

Bartram, William
  1928 *Travels of William Bartram* [1791]. Dover, New York.

Basgall, Mark E.
  1987 Resource Intensification among Hunter-Gatherers: Acorn Economies in Prehistoric California. *Research in Economic Anthropology* 9:21–52.

Bayman, James H.
  2001 The Hohokam of Southwest North America. *Journal of World Prehistory* 15:257–311.

Benson, Larry
  2009 Cahokia's Boom and Bust in the Context of Climate Change. *American Antiquity* 74:467–483.

Benson, Larry, and Michael S. Berry
  2009 Climate Change and Cultural Response in the Prehistoric American Southwest. *Kiva* 75:87–117.

Benson, Larry, Linda Cordell, Kirk Vincent, Howard Taylor, John Stein, G. Lang Farmer, and Kiyoto Futa
  2003 Ancient Maize from Chacoan Great Houses: Where Was It Grown? *Proceedings of the National Academy of Sciences* 100:13111–13115.

Benson, Larry, Kenneth Petersen, and John Stein
  2007 Anasazi (Pre-Columbian Native American) Migrations during the Middle-12th and Late-13th Centuries: Were They Drought-Induced? *Climatic Change* 83:187–213.

Bettinger, Robert L.
  1993 Doing Great Basin Archaeology Recently: Coping with Variability. *Journal of Archaeological Research* 1:43–66.
  2001 Holocene Hunter-Gatherers. In *Archaeology at the Millennium: A Sourcebook*, edited by Gary M. Feinman and T. Douglas Price, pp. 137–198. Springer, New York.
  2009 *Hunter-Gatherer Foraging: Five Simple Models*. Eliot Werner, Clinton Corners, New York.

Bettinger, Robert L., and Eric Wohlgemuth
  2011 Archaeological and Ethnographic Evidence for Indigenous Plant Use in California. In *The Subsistence Economies of Indigenous North American Societies: A Handbook*, edited by Bruce D. Smith, pp. 113–130. Smithsonian Institution Scholarly Press, Washington, DC.

Bird, Douglas, and James O'Connell
  2012 Human Behavioral Ecology. In *Archaeological Theory Today*, edited by Ian Hodder, pp. 37–61. Polity Press, Cambridge.

Blackman, Benjamin K., Moira Scascitelli, Nolan C. Kane, Harry H. Luton, David A. Rasmussen, Robert A. Bye, David L. Lentz, and Loren H. Rieseberg
  2011 Sunflower Domestication Alleles Support Single Domestication Center in Eastern North America. *Proceedings of the National Academy of Sciences* 108:14360–14365.

Blake, Leonard W.
  1981 Early Acceptance of Watermelon by Indians of the United States. *Journal of Ethnobiology* 1:193–199.

Blitz, John H.
  1993 Big Pots for Big Shots—Feasting and Storage in a Mississippian Community. *American Antiquity* 58:80–96.

Bohrer, Vorsila L.
  1992 New Life from Ashes II: A Tale of Burnt Brush. *Desert Plants* 10:122–125.

Bonzani, Renee M., and Augusto Oyuela-Caycedo
  2006 The Gift of the Variation and Dispersion of Maize: Social and Technological Context in Amerindian Societies. In *Histories of Maize: Multidisciplinary Approaches to the Prehistory, Linguistics, Biogeography, Domestication, and Evolution of Maize*, edited by John E. Staller, Robert H. Tykot, and Bruce F. Benz, pp. 343–356. Left Coast Press, Walnut Creek, California.

Boone, James L.
  2002 Subsistence Strategies and Early Human Population History: An Evolutionary Ecological Perspective. *World Archaeology* 34:6–25.

Bowles, Samuel, and Jung-Kyoo Choi
  2013 Coevolution of Farming and Private Property during the Early Holocene. *Proceedings of the National Academy of Sciences* 110:8830–8835.

Boyd, M., C. Surette, and B. A. Nicholson
  2006 Archaeobotanical Evidence of Prehistoric Maize (*Zea mays*) Consumption at the Northern Edge of the Great Plains. *Journal of Archaeological Science* 33:1129–1140.

Boyd, M., T. Varney, C. Surette, and J. Surette
  2008 Reassessing the Northern Limit of Maize Consumption in North America: Stable Isotope, Plant Microfossil, and Trace Element Content of Carbonized Food Residue. *Journal of Archaeological Science* 35:2545–2556.

Boyd, Matthew, and Clarence Surette
  2010 Northernmost Precontact Maize in North America. *American Antiquity* 75:117–133.

Boyd, Robert, and Peter J. Richerson
  1985 *Culture and the Evolutionary Process*. University of Chicago Press, Chicago.

Boyd, Robert, Peter J. Richerson, and Joseph Henrich
   2011   The Cultural Niche: Why Social Learning Is Essential for Human Adaptation. *Proceedings of the National Academy of Sciences* 108 (Supplement 2):10918–10925.

Braun, David P.
   1987   Coevolution of Sedentism, Pottery Technology, and Horticulture in the Central Midwest, 200 B.C. To A.D. 600. In *Emergent Horticultural Economies of the Eastern Woodlands*, edited by William F. Keegan, pp. 153–182. Occasional Paper No. 7. Center for Archaeological Investigations, Southern Illinois University at Carbondale, Carbondale.

Braund, Kathryn E. Holland
   1993   *Deerskins & Duffels: Creek Indian Trade with Anglo-America, 1685–1815*. University of Nebraska Press, Lincoln.

Briggs, Rachel V.
   2016   The Civil Cooking Pot: Hominy and the Mississippian Standard Jar in the Black Warrior Valley, Alabama. *American Antiquity* 81:316–332.

Bronk Ramsey, Christopher
   2009   Bayesian Analysis of Radiocarbon Dates. *Radiocarbon* 51:337–360.

Broughton, Jack M., Michael D. Cannon, and Eric J. Bartelink
   2010   Evolutionary Ecology, Resource Depression, and Niche Construction Theory: Applications to Central California Hunter-Gatherers and Mimbres-Mogollon Agriculturalists. *Journal of Archaeological Method and Theory* 17:371–421.

Brown, Cecil H., Charles R. Clement, Patience Epps, Eike Luedeling, and Søren Wichmann
   2014   The Paleobiolinguistics of the Common Bean (*Phaseolus vulgaris* L.). *Ethnobiology Letters* 5:104.

Brown, Ian W.
   1994   Recent Trends in the Archaeology of the Southeastern United States. *Journal of Archaeological Research* 2:45–112.

Bryant, Vaughn M.
   1974   The Role of Coprolite Analysis in Archaeology. *Bulletin of the Texas Archaeological Society* 45:1–28.

Buikstra, Jane E., Lyle W. Konigsberg, and Jill Bullington
   1986   Fertility and the Development of Agriculture in the Prehistoric Midwest. *American Antiquity* 51:528–546.

Butler, Virginia L., and Sarah K. Campbell
   2004   Resource Intensification and Resource Depression in the Pacific Northwest of North America: A Zooarchaeological Review. *Journal of World Prehistory* 18:327–405.

Campbell, Sarah K., and Virginia L. Butler
   2011   Prehistoric Native American Use of Animals on the Northwest Coast and Plateau. In *The Subsistence Economies of Indigenous North American Societies: A Handbook*, edited by Bruce D. Smith, pp. 83–111. Smithsonian Institution Scholarly Press, Washington, DC.

Campbell, T. N.
   1959   Choctaw Subsistence: Ethnographic Notes from the Lincecum Manuscript. *Florida Anthropologist* 12:9–24.

Cane, Scott
    1989 Australian Aboriginal Seed Grinding and Its Archaeological Record: A Case Study from the Western Desert. In *Foraging and Farming: The Evolution of Plant Exploitation*, edited by David R. Harris and Gordon C. Hillman, pp. 99–119. Unwin Hyman, London.

Cannon, Michael D.
    2000 Large Mammal Relative Abundance in Pithouse and Pueblo Period Archaeofaunas from Southwestern New Mexico: Resource Depression among the Mimbres-Mogollon? *Journal of Anthropological Archaeology* 19:317–347.
    2003 A Model of Central Place Forager Prey Choice and an Application to Faunal Remains from the Mimbres Valley, New Mexico. *Journal of Anthropological Archaeology* 22:1–25.

Carney, Judith A.
    1993 From Hands to Tutors: African Expertise in the South Carolina Rice Economy. *Agricultural History* 67:1–30.

Carney, Judith A., and Richard Porcher
    1993 Geographies of the Past: Rice, Slaves and Technological Transfer in South Carolina. *Southeastern Geographer* 33:127–147.

Chambers, Fiona Hamersley, and Nancy J. Turner
    2011 Plant Use by Northwest Coast and Plateau Indigenous Peoples. In *The Subsistence Economies of Indigenous North American Societies: A Handbook*, edited by Bruce D. Smith, pp. 65–82. Smithsonian Institution Scholarly Press, Washington, DC.

Chapman, Jefferson, and Gary Crites
    1987 Evidence for Early Maize (*Zea mays*) from the Icehouse Bottom Site, Tennessee. *American Antiquity* 52:352–354.

Chapman, Jefferson, Paul A. Delcourt, Patricia A. Cridlebaugh, Andrea B. Shea, and Hazel R. Delcourt
    1982 Man-Land Interaction: 10,000 Years of American Indian Impact on Native Ecosystems in the Lower Little Tennessee River Valley. *Southeastern Archaeology* 2:115–121.

Chapman, Jefferson, and Andrea B. Shea
    1981 The Archaeobotanical Record: Early Archaic Period to Contact in the Lower Little Tennessee River Valley. *Tennessee Anthropologist* 6:61–84.

Chilton, Elizabeth S.
    2008 So Little Maize, So Much Time: Understanding Maize Adoption in New England. In *Current Northeast Paleoethnobotany II*, edited by John P. Hart, pp. 53–59. New York State Museum Bulletin 512, Albany.

Chisholm, Brian S., D. Erle Nelson, and Henry P. Schwarcz
    1982 Stable-Carbon Isotope Ratios as a Measure of Marine versus Terrestrial Protein in Ancient Diets. *Science* 216:1131–1132.

Chisholm, Brian, and R. G. Matson
    1994 Carbon and Nitrogen Isotopic Evidence on Basketmaker II Diet at Cedar Mesa, Utah. *Kiva* 60:239–255.

Chomko, Stephen, and Gary W. Crawford
    1978 Plant Husbandry in Eastern North America: New Evidence for Its Development. *American Antiquity* 43:405–408.

Clifford, Walter Allen, IV
    2012 Paleoethnobotanical Analysis of 38bk1633. Master's thesis, Department of Anthropology, University of South Carolina, Columbia.

Coe, Sophie D.
    1994 *America's First Cuisines*. University of Texas Press, Austin.

Coles, Nathan D., Michael D. McMullen, Peter J. Balint-Kurti, Richard C. Pratt, and James B. Holland
    2010 Genetic Control of Photoperiod Sensitivity in Maize Revealed by Joint Multiple Population Analysis. *Genetics* 184:799–812.

Coltrain, Joan Brenner, Joel C. Janetski, and Shawn W. Carlyle
    2007 The Stable- and Radio-Isotope Chemistry of Western Basketmaker Burials: Implications for Early Puebloan Diets and Origins. *American Antiquity* 72:301–321.

Coltrain, Joan Brenner, and Steven W. Leavitt
    2002 Climate and Diet in Fremont Prehistory: Economic Variability and Abandonment of Maize Agriculture in the Great Salt Lake Basin. *American Antiquity* 67:453–485.

Commission for Environmental Cooperation
    1997 *Ecological Regions of North America: Toward a Common Perspective*. Electronic document, http://www3.cec.org/islandora/en/item/1701-ecological-regions-north-america-toward-common-perspective, accessed June 1, 2016.
    2011 North American Terrestrial Ecoregions—Level III. In *North American Environmental Atlas*. Electronic document, http://www3.cec.org/islandora/en/item/10415-north-american-terrestrial-ecoregionslevel-iii, accessed June 1, 2016.

Conard, Nicholas, David L. Asch, Nancy B. Asch, David Elmore, Harry Gove, Meyer Rubin, James A. Brown, Michael D. Wiant, Kenneth B. Farnsworth, and Thomas G. Cook
    1983 Prehistoric Horticulture in Illinois: Accelerator Radiocarbon Dating of the Evidence. *Nature* 308:443–446.

Cook, Robert A., and T. Douglas Price
    2015 Maize, Mounds, and the Movement of People: Isotope Analysis of a Mississippian/Fort Ancient Region. *Journal of Archaeological Science* 61:112–128.

Cook, Robert A., and Mark R. Schurr
    2009 Eating between the Lines: Mississippian Migration and Stable Carbon Isotope Variation in Fort Ancient Populations. *American Anthropologist* 111:344–359.

Cordell, Linda, and Maxine E. McBrinn
    2012 *Archaeology of the Southwest*. 3rd ed. Routledge, London.

Cowan, C. Wesley
    1978a The Prehistoric Use and Distribution of Maygrass in Eastern North America: Cultural and Phytogeographical Implications. In *The Nature and Status of Ethnobotany*, edited by Richard I. Ford, pp. 263–288. Museum of Anthropology, University of Michigan, Ann Arbor.

1978b Seasonal Nutritional Stress in a Late Woodland Population: Suggestions from Some Eastern Kentucky Coprolites. *Tennessee Anthropologist* 3:117–128.

1985a From Foraging to Incipient Food Production: Subsistence Change and Continuity on the Cumberland Plateau of Eastern Kentucky. PhD dissertation, Department of Anthropology, University of Michigan, Ann Arbor.

1985b Understanding the Evolution of Plant Husbandry in Eastern North America: Lessons from Botany, Ethnography, and Archaeology. In *Prehistoric Food Production in North America*, edited by Richard I. Ford, pp. 205–243. Museum of Anthropology, University of Michigan, Ann Arbor.

1997 Evolutionary Changes Associated with the Domestication of *Cucurbita pepo*: Evidence from Eastern Kentucky. In *People, Plants, and Landscapes: Studies in Paleoethnobotany*, edited by Kristen J. Gremillion, pp. 63–85. University of Alabama Press, Tuscaloosa.

Crawford, Gary W.
2011 People and Plant Interactions in the Northeast. In *The Subsistence Economies of Indigenous North American Societies: A Handbook*, edited by Bruce D. Smith, pp. 431–448. Smithsonian Institution Scholarly Press, Washington, DC.

Crawford, Gary W., and David G. Smith
2003 Paleoethnobotany in the Northeast. In *People and Plants in Ancient Eastern North America*, edited by Paul E. Minnis, pp. 172–257. Smithsonian Books, Washington, DC.

Crawford, Gary W., David G. Smith, and Vandy E. Bowyer
1997 Dating the Entry of Corn (*Zea mays*) into the Lower Great Lakes Region. *American Antiquity* 62:112–120.

Crawford, Gary W., David G. Smith, Joseph R. Desloges, and Anthony M. Davis
1998 Floodplains and Agriculture Origins: A Case Study in South-Central Ontario, Canada. *Journal of Field Archaeology* 25:123–137.

Crites, Gary D.
1978 Plant Food Utilization Patterns during the Middle Woodland Owl Hollow Phase in Tennessee: A Preliminary Report. *Tennessee Anthropologist* 3:80–92.

1987 Middle and Late Holocene Paleoethnobotany of the Hayes Site (40ml139): Evidence from Unit 990n918e. *Midcontinental Journal of Archaeology* 12:3–15.

1991 Investigations into Early Plant Domesticates and Food Production in Middle Tennessee: A Status Report. *Tennessee Anthropologist* 16:69–87.

1993 Domesticated Sunflower in Fifth Millennium B.P. Temporal Context: New Evidence from Middle Tennessee. *American Antiquity* 58:146–148.

Crites, Gary D., and R. Dale Terry
1984 Nutritive Value of Maygrass, *Phalaris caroliniana*. *Economic Botany* 38:114–120.

Cronon, William
1983 *Changes in the Land: Indians, Colonists, and the Ecology of New England*. Hill and Wang, New York.

Crosby, Alfred
1976 Virgin Soil Epidemics as a Factor in the Aboriginal Depopulation in America. *William and Mary Quarterly* 33:289–299.

1986 *Ecological Imperialism: The Biological Expansion of Europe, 900–1900.* Cambridge University Press, Cambridge.

Crothers, George

2008 From Foraging to Farming: The Emergence of Exclusive Property Rights in Kentucky Prehistory. In *Economies and the Transformation of Landscape*, edited by Lisa Cliggett and Christopher A. Poole, pp. 127–147. Society for Economic Anthropology Monographs 25. AltaMira Press, Lanham, Maryland.

Crown, Patricia L. and W. Jeffrey Hurst

2009 Evidence of Cacao Use in the Prehispanic American Southwest. *Proceedings of the National Academy of Sciences* 106:2110–2113.

Damp, Jonathan E., Stephen A. Hall, and Susan J. Smith

2002 Early Irrigation on the Colorado Plateau near Zuni Pueblo, New Mexico. *American Antiquity* 67:665–676.

Day, Gordon

1953 The Indian as an Ecological Factor in the Northeastern Forest. *Ecology* 34:329–336.

Dean, Rebecca M.

2005 Site-Use Intensity, Cultural Modification of the Environment, and the Development of Agricultural Communities in Southern Arizona. *American Antiquity* 70:403–431.

DeBoer, Warren R.

2004 Little Bighorn on the Scioto: The Rocky Mountain Connection to Ohio Hopewell. *American Antiquity* 69:85–107.

Decker, Deena S.

1988 Origin(s), Evolution, and Systematics of *Cucurbita pepo* (Cucurbitaceae). *Economic Botany* 42:4–15.

Decker-Walters, Deena S., Terrence W. Walters, C. Wesley Cowan, and Bruce D. Smith

1993 Isozymic Characterization of Wild Populations of *Cucurbita pepo*. *Journal of Ethnobiology* 13:55–74.

Delcourt, Hazel R.

2008 *Prehistoric Native Americans and Ecological Change: Human Ecosystems in Eastern North America since the Pleistocene.* Cambridge University Press, Cambridge.

Delcourt, Paul A., and Hazel R. Delcourt

1997 *Report of Paleoecological Investigations: Cliff Palace Pond, Jackson County, Kentucky, in the Daniel Boone National Forest.* Report submitted to the USDA Forest Service, Daniel Boone National Forest, Winchester, Kentucky. Contract Order No. 43-531A-6-0389. Winchester, Kentucky.

Delcourt, Paul A., Hazel R. Delcourt, Patricia A. Cridlebaugh, and Jefferson Chapman

1986 Holocene Ethnobotanical and Paleoecological Record of Human Impact on Vegetation in the Little Tennessee River Valley, Tennessee. *Quaternary Research* 25:330–349.

Delcourt, Paul A., Hazel Delcourt, Cecil R. Ison, William Sharp, and Kristen J. Gremillion
  1998 Prehistoric Human Use of Fire, the Eastern Agricultural Complex, and Appalachian Oak-Chestnut Forests: Paleoecology of Cliff Palace Pond, Kentucky. *American Antiquity* 63:263–278.
Deur, Douglas
  2005 Tending the Garden, Making the Soil: Northwest Coast Estuarine Gardens as Engineered Environments. In *Keeping It Living: Traditions of Plant Use and Cultivation on the Northwest Coast of North America*, edited by Douglas Deur and Nancy J. Turner, pp. 296–330. University of Washington Press, Seattle.
Deur, Douglas, and Nancy J. Turner
  2005 Introduction: Reconstructing Indigenous Resource Management, Reconstructing the History of an Idea. In *Keeping It Living: Traditions of Plant Use and Cultivation on the Northwest Coast of North America*, edited by Douglas Deur and Nancy J. Turner, pp. 3–34. University of Washington Press, Seattle.
deWet, J. M. J.
  1975 Evolutionary Dynamics of Cereal Domestication. *Bulletin of the Torrey Botanical Club* 102:307–312.
deWet, J. M. J., and J. R. Harlan
  1975 Weeds and Domesticates: Evolution in a Man-Made Habitat. *Economic Botany* 29:99–107.
Dezendorf, Caroline
  2013 The Effects of Food Processing on the Archaeological Visibility of Maize: An Experimental Study of Carbonization of Lime Treated Maize Kernels. *Ethnobiology Letters* 4:12–20.
Diehl, Michael
  1996 The Intensity of Maize Processing and Production in Upland Mogollon Pithouse Villages, A.D. 200–1000. *American Antiquity* 61:102–115.
  1997 Rational Behavior, the Adoption of Agriculture, and the Organization of Subsistence during the Late Archaic Period in the Greater Tucson Basin. In *Rediscovering Darwin: Evolutionary Theory and Archeological Explanations*, edited by C. Michael Barton and Geoffrey A. Clark, pp. 251–265. Archeological Papers No. 7. American Anthropological Association, Washington, DC.
  2005 Morphological Observations on Recently Recovered Early Agricultural Period Maize Cob Fragments from Southern Arizona. *American Antiquity* 70:361–375.
  2012 Subsistence during the Pithouse Periods. In *Southwestern Pithouse Communities, AD 200–900*, edited by Lisa C. Young and Sarah A. Herr, pp. 14–33. University of Arizona Press, Tucson.
Diehl, Michael, and Owen K. Davis
  2016 The Short, Unhappy Use Lives of Early Agricultural Period "Food Storage" Pits at the Las Capas Site, Southern Arizona. *American Antiquity* 81:333–344.
Diehl, Michael, and Jennifer A. Waters
  2006 Aspects of Optimization and Risk during the Early Agricultural Period in Southwestern Arizona. In *Behavioral Ecology and the Transition to Agriculture*, edited

by Douglas J. Kennett and Bruce Winterhalder, pp. 63–86. University of California Press, Berkeley.

Doebley, John F.
1984 "Seeds" of Wild Grasses: A Major Food of Southwestern Indians. *Economic Botany* 38:52–64.

Doebley, John F., Major M. Goodman, and Charles W. Stuber
1986 Exceptional Genetic Divergence of Northern Flint Corn. *American Journal of Botany* 73:64–69.

Dominguez, Steven
2002 Optimal Gardening Strategies: Maximizing the Input and Retention of Water in Prehistoric Gridded Fields in North Central New Mexico. *World Archaeology* 34:131–163.

Doolittle, William E.
2000 *Cultivated Landscapes of Native North America*. Oxford University Press, Oxford.
2004 Permanent vs. Shifting Cultivation in the Eastern Woodlands of North America Prior to European Contact. *Agriculture and Human Values* 21:181–189.

Doolittle, William E., and Jonathan B. Mabry
2006 Environmental Mosaics, Agricultural Diversity, and the Evolutionary Adoption of Maize in the American Southwest. In *Histories of Maize: Multidisciplinary Approaches to the Prehistory, Linguistics, Biogeography, Domestication, and Evolution of Maize*, edited by John E. Staller, Robert H. Tykot, and Bruce F. Benz, pp. 109–121. Left Coast Press, Walnut Creek, California.

Dorshow, Wetherbee Tryan
2012 Modeling Agricultural Potential in Chaco Canyon during the Bonito Phase: A Predictive Geospatial Approach. *Journal of Archaeological Science* 39:2098–2115.

Dunavan, Sandra L.
1993 Reanalysis of Seed Crops from Emge: New Implications for Late Woodland Subsistence-Settlement Systems. In *Foraging and Farming in the Eastern Woodlands*, edited by C. Margaret Scarry, pp. 98–114. University Press of Florida, Gainesville.

Dunavan, Sandra L., and Volney Jones
2011 Tobacco and Smoking in Native North America. In *The Subsistence Economies of Indigenous North American Societies: A Handbook*, edited by Bruce D. Smith, pp. 517–524. Smithsonian Institution Scholarly Press, Washington, DC.

Dyer, James M.
2006 Revisiting the Deciduous Forests of Eastern North America. *BioScience* 56:341–352.

Eerkens, Jelmer W.
2003 Sedentism, Storage, and the Intensification of Small Seeds: Prehistoric Developments in Owens Valley, California. *North American Archaeologist* 24:281–309.
2004 Privatization, Small-Seed Intensification, and the Origins of Pottery in the Western Great Basin. *American Antiquity* 69:653–670.

Eerkens, Jelmer W., Devin L. Snyder, and Nicole A. Reich
  2008 Rock-Ring Features on the Shores of Owens Lake and Implications for Prehistoric Geophyte Processing and Storage. *Proceedings of the Society for California Archaeology* 21:180–184.
Elvas, Gentleman of
  1993 The True Relation of the Hardships Suffered by Governor Hernando De Soto. In *The De Soto Chronicles: The Expedition of Hernando de Soto to North America in 1539–1543*, edited by Lawrence A. Clayton, Vernon James Knight Jr., and Edward C. Moore, pp. 19–220. Translated by J. A. Robertson. University of Alabama Press, Tuscaloosa.
Emerson, T. E., Kristin M. Hedman, and Mary L. Simon
  2005 Marginal Horticulturalists or Maize Agriculturalists? Archaeobotanical, Paleopathological, and Isotopic Evidence Relating to Langford Tradition Maize Consumption. *Midcontinental Journal of Archaeology* 30:67–118.

Faulkner, Charles T.
  1991 Prehistoric Diet and Parasitic Infection in Tennessee: Evidence from the Analysis of Desiccated Human Paleofeces. *American Antiquity* 56:687–700.
Fish, Suzanne K.
  2004 Corn, Crops, and Cultivation in the North American Southwest. In *People and Plants in Ancient Western North America*, edited by Paul E. Minnis, pp. 115–166. Smithsonian Institution Press, Washington, DC.
Fish, Suzanne K., and Paul R. Fish
  1992 Prehistoric Landscapes of the Sonoran Desert Hohokam. *Population and Environment* 13:269–283.
Fisher, Robin
  1996 The Northwest from the Beginning of Trade with Europeans to the 1880s. In *The Cambridge History of the Native Peoples of the Americas*, edited by Bruce G. Trigger and Wilcomb E. Washburn, pp. 117–182. Cambridge University Press, Cambridge.
Ford, Richard I.
  1981 Gardening and Farming Before A.D. 1000: Patterns of Prehistoric Cultivation North of Mexico. *Journal of Ethnobiology* 1:6–27.
Ford, Richard I., and Roxanne Swentzell
  2015 Precontact Agriculture in Northern New Mexico. In *Traditional Arid Lands Agriculture: Understanding the Past for the Future*, edited by Scott E. Ingram and Robert C. Hunt, pp. 330–357. University of Arizona Press, Tucson.
Foster, H. Thomas, II
  2003 Dynamic Optimization of Horticulture among the Muscogee Creek Indians of the Southeastern United States. *Journal of Anthropological Archaeology* 22:411–424.
  2009 Cultural Continuity and Variable Adaptation of the Historic Period Muscogee of the Southeastern United States. *North American Archaeologist* 30:259–290.
  2010 Risk Management among Native American Horticulturalists of the Southeastern United States (1715–1825). *Journal of Anthropological Research* 66:69–96.

Fowler, Catherine S.
   2000 "We Live by Them": Native Knowledge of Biodiversity in the Great Basin of Western North America. In *Biodiversity and Native America*, edited by Paul E. Minnis and Wayne J. Elisens, pp. 99–132. University of Oklahoma Press, Norman.

Fowler, Catherine S., and David E. Rhode
   2011 Plant Foods and Foodways among the Great Basin's Indigenous Peoples. In *The Subsistence Economies of Indigenous North American Societies: A Handbook*, edited by Bruce D. Smith, pp. 233–270. Smithsonian Institution Scholarly Press, Washington, DC.

Fowler, Loretta, Bruce G. Trigger, and Wilcomb E. Washburn
   1996 The Great Plains from the Arrival of the Horse to 1885. In *The Cambridge History of the Native Peoples of the Americas*, edited by Bruce G. Trigger and Wilcomb E. Washburn, pp. 1–56. Cambridge University Press, Cambridge.

Fritz, Gayle J.
   1984a Identification of Cultigen Amaranth and Chenopod from Rockshelter Sites in Northwest Arkansas. *American Antiquity* 49:558–572.
   1984b Prehistoric Ozark Agriculture: The University of Arkansas Rockshelter Collections. PhD dissertation, Department of Anthropology, University of North Carolina, Chapel Hill.
   1990 Multiple Pathways to Farming in Precontact Eastern North America. *Journal of World Prehistory* 4:387–435.
   1993 Early and Middle Woodland Period Paleoethnobotany. In *Foraging and Farming in the Eastern Woodlands*, edited by C. Margaret Scarry, pp. 39–56. University Press of Florida, Gainesville.
   1994 In Color and in Time: Prehistoric Ozark Agriculture. In *Agricultural Origins and Development in the Midcontinent*, edited by William Green, pp. 105–126. Office of the State Archaeologist, University of Iowa, Iowa City.
   1997 A Three-Thousand-Year-Old Cache of Crop Seeds from Marble Bluff, Arkansas. In *People, Plants, and Landscapes: Studies in Paleoethnobotany*, edited by Kristen J. Gremillion, pp. 42–62. University of Alabama Press, Tuscaloosa.
   1999 Gender and the Early Cultivation of Gourds in Eastern North America. *American Antiquity* 64:417–430.
   2000 Levels of Native Biodiversity in Eastern North America. In *Biodiversity and Native America*, edited by Paul E. Minnis and Wayne J. Elisens, pp. 223–247. University of Oklahoma Press, Norman.
   2011 The Role of "Tropical Crops" in Early North American Agriculture. In *The Subsistence Economies of Indigenous North American Societies: A Handbook*, edited by Bruce D. Smith, pp. 503–516. Smithsonian Institution Scholarly Press, Washington, DC.
   2014 Maygrass (*Phalaris caroliniana* Walt.): Its Role and Significance in Native Eastern North American Agriculture. In *New Lives for Ancient and Extinct Crops*, edited by Paul E. Minnis, pp. 12–43. University of Arizona Press, Tucson.

Fritz, Gayle J., and Tristram R. Kidder
   1993 Recent Investigations into Prehistoric Agriculture in the Lower Mississippi Valley. *Southeastern Archaeology* 12:1–14.

Fritz, Gayle J., and Bruce D. Smith
　1988　Old Collections and New Technology: Documenting the Domestication of *Chenopodium* in Eastern North America. *Midcontinental Journal of Archaeology* 13:3–27.

Gallagher, James P., and Constance Arzigian
　1994　A New Perspective on Late Prehistoric Agricultural Intensification in the Upper Mississippi River Valley. In *Agricultural Origins and Development in the Midcontinent*, edited by William Green, pp. 171–184. Report 19. Office of the State Archaeologist, University of Iowa, Iowa City.

Gardner, Paul S.
　1997　The Ecological Structure and Behavioral Implications of Mast Exploitation Strategies. In *People, Plants, and Landscapes: Studies in Paleoethnobotany*, edited by Kristen J. Gremillion, pp. 161–178. University of Alabama Press, Tuscaloosa.

Ghadim, Amir K. Abadi, and David J. Pannell
　1999　A Conceptual Framework of Adoption of an Agricultural Innovation. *Agricultural Economics* 21:145–154.

Gigerenzer, Gerd, and Daniel G. Goldstein
　1996　Reasoning the Fast and Frugal Way: Models of Bounded Rationality. *Psychological Review* 103:650–669.

Gilman, Patricia A
　1987　Architecture as Artifact: Pit Structures and Pueblos in the American Southwest. *American Antiquity* 52:538–564.

Gilman, Patricia A., Elizabeth M. Toney, and Nicholas H. Beale
　2013　A Historical Ecological Perspective on Early Agriculture in the North American Southwest and Northwest Mexico. In *The Archaeology and Historical Ecology of Small Scale Economies*, edited by Victor D. Thompson and James C. Waggoner, pp. 96–118. Univeristy Press of Florida, Gainesville.

Green, Michael D., Bruce G. Trigger, and Wilcomb E. Washburn
　1996　The Expansion of European Colonization to the Mississippi Valley, 1780–1880. In *The Cambridge History of the Native Peoples of the Americas*, edited by Bruce G. Trigger and Wilcomb E. Washburn, pp. 461–538. Cambridge University Press, Cambridge.

Greenlee, Diana M.
　2006　Dietary Variation and Prehistoric Maize Farming in the Middle Ohio Valley. In *Histories of Maize: Multidisciplinary Approaches to the Prehistory, Linguistics, Biogeography, Domestication, and Evolution of Maize*, edited by John E. Staller, Robert H. Tykot, and Bruce F. Benz, pp. 215–233. Left Coast Press, Walnut Creek, California.

Gremillion, Kristen J.
　1993a　Adoption of Old World Crops and Processes of Cultural Change in the Historic Southeast. *Southeastern Archaeology* 12:15–20.
　1993b　Crop and Weed in Prehistoric Eastern North America: The *Chenopodium* Example. *American Antiquity* 58:496–509.
　1993c　The Evolution of Seed Morphology in Domesticated *Chenopodium*: An Archaeological Case Study. *Journal of Ethnobiology* 13:149–169.

1993d Plant Husbandry at the Archaic/Woodland Transition: Evidence from the Cold Oak Shelter, Kentucky. *Midcontinental Journal of Archaeology* 18:161–189.

1994 Evidence of Plant Domestication from Kentucky Caves and Rockshelters. In *Agricultural Origins and Development in the Midcontinent*, edited by William Green, pp. 87–104. Office of the State Archaeologist, University of Iowa, Iowa City.

1995a Archaeological and Paleoethnobotanical Investigations at the Cold Oak Shelter, Kentucky. Report submitted to the National Geographic Society, Grant 5226–94. Copy on file at Department of Anthropology, Ohio State University, Columbus.

1995b Comparative Paleoethnobotany of Three Native Southeastern Communities of the Historic Period. *Southeastern Archaeology* 14:1–16.

1996a Diffusion and Adoption of Crops in Evolutionary Perspective. *Journal of Anthropological Archaeology* 15:183–204.

1996b Early Agricultural Diet in Eastern North America: Evidence from Two Kentucky Rockshelters. *American Antiquity* 61:520–536.

1997 New Perspectives on the Paleoethnobotany of the Newt Kash Shelter. In *People, Plants, and Landscapes: Studies in Paleoethnobotany*, edited by Kristen J. Gremillion, pp. 23–41. University of Alabama Press, Tuscaloosa.

2002a The Development and Dispersal of Agricultural Systems in the Woodland Period Southeast. In *The Woodland Southeast*, edited by David G. Anderson and Robert C. Mainfort Jr., pp. 483–501. University of Alabama Press, Tuscaloosa.

2002b Human Ecology at the Edge of History. In *Between Contacts and Colonists: Protohistoric Archaeology in the Southeastern United States*, edited by Mark Rees and Cameron Wesson, pp. 12–31. University of Alabama Press, Tuscaloosa.

2004 Seed Processing and the Origins of Food Production in Eastern North America. *American Antiquity* 69:215–234.

2008 From Dripline to Deep Cave: On Sheltered Sites as Archaeobotanical Contexts. In *Cave Archaeology of the Eastern Woodlands: Essays in Honor of Patty Jo Watson*, edited by David H. Dye, pp. 117–126. University of Tennessee Press, Knoxville.

2014 Goosefoot (*Chenopodium*). In *New Lives for Ancient and Extinct Crops*, edited by Paul E. Minnis, pp. 44–64. University of Arizona Press, Tucson.

2015 Prehistoric Upland Farming, Fuelwood, and Forest Composition on the Cumberland Plateau, Kentucky, USA. *Journal of Ethnobiology* 35:60–84.

Gremillion, Kristen J., Loukas Barton, and Dolores R. Piperno

2014 Particularism and the Retreat from Theory in the Archaeology of Agricultural Origins. *Proceedings of the National Academy of Sciences* 111:6171–6177.

Gremillion, Kristen J., and Cecil R. Ison

1993 Terminal Archaic and Early Woodland Plant Utilization at the Cold Oak Shelter. In *Upland Archaeology in the East: Symposium IV*, edited by Michael D. Barber and Eugene B. Barfield, pp. 121–132. USDA Forest Service, Southern Region, Atlanta, Georgia.

Gremillion, Kristen J., and Kristin D. Sobolik

1996 Dietary Variability among Prehistoric Forager-Farmers of Eastern North America. *Current Anthropology* 37:529–539.

Gremillion, Kristen J., Jason Windingstad, and Sarah S. Sherwood
  2008 Forest Opening, Habitat Use, and Food Production on the Cumberland Plateau, Kentucky: Adaptive Flexibility in Marginal Settings. *American Antiquity* 73:387–411.
Grimstead, Deanna N., Jay Quade, and M. C. Stiner
  2016 Isotopic Evidence for Long-Distance Mammal Procurement, Chaco Canyon, New Mexico, USA. *Geoarchaeology* 31:335–354.

Hammerstedt, Scott W., and Erin R. Hughes
  2015 Mill Creek Chert Hoes and Prairie Soils: Implications for Cahokian Production and Expansion. *Midcontinental Journal of Archaeology* 40:149–165.
Hammett, Julia E.
  2000 Ethnohistory of Aboriginal Landscapes in the Southeastern United States. In *Biodiversity and Native America*, edited by Paul E. Minnis and Wayne J. Elisens, pp. 248–299. University of Oklahoma Press, Norman.
Hard, Robert J., Raymond P. Mauldin, and Gerry R. Raymond
  1996 Mano Size, Stable Carbon Isotope Ratios, and Macrobotanical Remains as Multiple Lines of Evidence of Maize Dependence in the American Southwest. *Journal of Archaeological Method and Theory* 3:253–318.
Hard, Robert J., and J. R. Roney
  2005 The Transition to Farming on the Rio Casa Grandes and in the Southern Jornada Mogollon Region. In *The Late Archaic across the Borderlands: From Foraging to Farming*, edited by Bradley J. Vierra, pp. 141–186. University of Texas Press, Austin.
Harlan, Jack R., and J. M. J. deWet
  1965 Some Thoughts about Weeds. *Economic Botany* 19:16–24.
Harlan, Jack R., J. M. J. deWet, and Glen E. Price
  1973 Comparative Evolution of Cereals. *Evolution* 27:311–325.
Harriot, Thomas
  1972 [1590] *A Briefe and True Report of the New Found Land of Virginia*. Dover, New York.
Harris, David R.
  1989 An Evolutionary Continuum of People-Plant Interaction. In *Foraging and Farming: The Evolution of Plant Exploitation*, edited by David R. Harris and Gordon C. Hillman, pp. 11–26. Unwin Hyman, London.
  2007 Agriculture, Cultivation, and Domestication: Exploring the Conceptual Framework of Early Food Production. In *Rethinking Agriculture: Archaeological and Ethnoarchaeological Perspectives*, edited by Tim Denham, Jose Iriarte, and Luc Vrydaghs, pp. 16–35. Left Coast Press, Walnut Creek, California.
Harris, Marvin
  1968 *The Rise of Anthropological Theory*. Thomas Crowell Company, New York.
Hart, John P.
  1990 Modeling Oneota Agricultural Production: A Cross-Cultural Evaluation. *Current Anthropology* 31:569–577.
  1999 Maize Agriculture Evolution in the Eastern Woodlands of North America: A Darwinian Perspective. *Journal of Archaeological Method and Theory* 6:137–179.

2008 Evolving the Three Sisters: The Changing Histories of Maize, Bean, and Squash in New York and the Greater Northeast. In *Current Northeast Paleoethnobotany II*, edited by John P. Hart, pp. 87–100. New York State Museum Bulletin 512, Albany.

2012 Pottery Wall Thinning as a Consequence of Increased Maize Processing: A Case Study from Central New York. *Journal of Archaeological Science* 39:3470–3474.

2014 A Critical Assessment of Current Approaches to Investigations of the Timing, Rate, and Adoption Trajectories of Domesticates in the Midwest and Great Lakes. In *Timing, Rate, and Adoption Trajectories of Domesticates in the Midwest and Great Lakes*, edited by Maria E. Raviele and William A. Lovis, pp. 161–174. Occasional Papers, No. 1. Midwestern Archaeological Conference, Inc.

Hart, John P., Hetty Jo Brumbach, and Robert Lusteck

2007 Extending the Phytolith Evidence for Early Maize (*Zea mays* ssp. *mays*) and Squash (*Cucurbita* sp.) in Central New York. *American Antiquity* 72:563–583.

Hart, John P., and William A. Lovis

2012 Reevaluating What We Know about the Histories of Maize in Northeastern North America: A Review of Current Evidence. *Journal of Archaeological Research* 21:175–216.

Hart, John P., and Nancy Asch Sidell

1997 Additional Evidence for Early Cucurbit Use in the Northern Eastern Woodlands East of the Allegheny Front. *American Antiquity* 62:523–537.

Hatley, M. Thomas

1989 The Three Lives of Keowee: Loss and Recovery in Eighteenth-Century Cherokee Villages. In *Powhatan's Mantle: Indians in the Colonial Southeast*, edited by Gregory A. Waselkov, Peter H. Wood, and M. Thomas Hatley, pp. 222–248. University of Nebraska Press, Lincoln.

Haury, Emil W.

1962 The Greater American Southwest. In *Courses Toward Urban Life: Some Archaeological Considerations of Some Cultural Alternates*, edited by Robert J. Braidwood and Gordon R. Willey, pp. 106–131. Viking Fund Publications in Anthropology No. 32, New York.

Hedman, Kristin M.

2006 Late Cahokian Subsistence and Health: Stable Isotope and Dental Evidence. *Southeastern Archaeology* 25:258–274.

Hegmon, Michelle

2002 Recent Issues in the Archaeology of the Mimbres Region of the North American Southwest. *Journal of Archaeological Research* 10:307–357.

Hegmon, Michelle, Margaret C. Nelson, and Susan M. Ruth

1998 Abandonment and Reorganization in the Mimbres Region of the American Southwest. *American Anthropologist* 100:1–15.

Henrich, Joseph

2001a Cultural Transmission and the Diffusion of Innovations: Adoption Dynamics Indicate that Biased Cultural Transmission Is the Predominate Force in Behavioral Change. *American Anthropologist* 103:992–1013.

2001b In Search of *Homo economicus*: Behavioral Experiments in 15 Small-Scale Societies. *AEA Papers and Proceedings* 91:73–78.

Henrich, Joseph, Robert Boyd, Samuel Bowles, Colin Camerer, Ernst Fehr, Herb Gintis, Richard McElreath, Michael Alvard, Abigail Barr, Jean Ensminger, Natalie Smith Henrich, Kim Hill, Francisco Gil-White, Michael Gurven, Frank W. Marlowe, John Q. Patton, and David Tracer
2005 "Economic Man" in Cross-Cultural Perspective: Behavioral Experiments in 15 Small-Scale Societies. *Behavioral and Brain Sciences* 28:795–815.

Henrich, Joseph, Robert Boyd, Maxime Derex, Michelle A. Kline, Alex Mesoudi, Michael Muthukrishna, Adam T. Powell, Stephen J. Shennan, and Mark G. Thomas
2016 Understanding Cumulative Cultural Evolution. *Proceedings of the National Academy of Sciences* 113:E6724–E6725.

Henrich, Joseph, and R. McElreath
2003 The Evolution of Cultural Evolution. *Evolutionary Anthropology* 12:123–135.

Herhahn, Cynthia L,. and J. Brett Hill
1998 Modeling Agricultural Production Strategies in the Northern Rio Grande Valley, New Mexico. *Human Ecology* 26:469–487.

Herring, Erin M., R. Scott Anderson, and George L. San Miguel
2014 Fire, Vegetation, and Ancestral Puebloans: A Sediment Record from Prater Canyon in Mesa Verde National Park, Colorado, USA. *The Holocene* 24:853–863.

Hildebrandt, William R., and Kimberley Carpenter
2011 Native Hunting Adaptations in California: Changing Patterns of Resource Use from the Early Holocene to European Contact. In *The Subsistence Economies of Indigenous North American Societies: A Handbook*, edited by Bruce D. Smith, pp. 131–146. Smithsonian Insitution Scholarly Press, Washington, DC.

Hill, Matthew E. Jr., J. Simon Bruder, Margaret E. Beck, and Bruce G. Phillips
2008 Mobile Horticulturalists in the Western Papagueria. *Kiva* 74:33–69.

Huckell, Bruce B.
1996 The Archaic Prehistory of the North American Southwest. *Journal of World Prehistory* 10:305–373.

Huckell, Lisa W., and Mollie S. Toll
2004 Wild Plant Use in the North American Southwest. In *People and Plants in Ancient Western North America*, edited by Paul E. Minnis, pp. 37–114. Smithsonian Institution Press, Washington, DC.

Hudson, Charles
1994 The Hernando De Soto Expedition, 1539–1543. In *The Forgotten Centuries: Indians and Europeans in the American South, 1521–1704*, edited by Charles Hudson and Carmen Chavez Tesser, pp. 74–103. University of Georgia Press, Athens.

Hutchinson, Dale L., Clark Spencer Larsen, Margaret J. Schoeninger, and Lynette Norr
1998 Regional Variation in the Pattern of Maize Adoption and Use in Florida and Georgia. *American Antiquity* 63:397–416.

Ison, Cecil R.
  1991 Prehistoric Upland Farming along the Cumberland Plateau. In *Studies in Kentucky Archaeology*, edited by Charles D. Hockensmith, pp. 1–10. Kentucky Heritage Council, Frankfort.

Jackson, H. Edwin, and Susan L. Scott
  2003 Patterns of Elite Faunal Utilization at Moundville, Alabama. *American Antiquity* 68:552–572.
Jaenicke-Després, Viviane, Ed S. Buckler, Bruce D. Smith, M. Thomas, P. Gilbert, Alan Cooper, John Doebley, and Svante Pääbo
  2003 Early Allelic Selection in Maize as Revealed by Ancient DNA. *Science* 302:1206–1208.
Johannessen, Sissel
  1984 Paleoethnobotany. In *American Bottom Archaeology*, edited by Charles J. Bareis and James W. Porter, pp. 197–214. University of Illinois Press, Urbana.
  1993 Farmers of the Late Woodland. In *Foraging and Farming in the Eastern Woodlands*, edited by C. Margaret Scarry, pp. 57–77. University Press of Florida, Gainesville.
Jones, Volney
  1936 The Vegetal Remains of Newt Kash Hollow Shelter. In *Rock Shelters in Menifee County, Kentucky*, edited by William S. Webb and William D. Funkhouser, pp. 147–165. Reports in Archaeology and Anthropology 3. University of Kentucky, Lexington.

Kantner, John
  2010 Implications of Human Behavioral Ecology for Understanding Complex Human Behavior: Resource Monopolization, Package Size, and Turquoise. *Journal of Anthropological Research* 66:231–257.
Katzenberg, M. Anne, Henry P. Schwarcz, Martin Knyf, and F. Jerome Melbye
  1995 Stable Isotope Evidence for Maize Horticulture and Paleodiet in Southern Ontario, Canada. *American Antiquity* 60:335–350.
Kay, Marvin, Frances B. King, and Christine K. Robinson
  1980 Cucurbits from Phillips Spring: New Evidence and Interpretations. *American Antiquity* 45:806–822.
Kelly, Dave, and Victoria L. Sork
  2002 Mast Seeding in Perennial Plants: Why, How, Where? *Annual Review of Ecology and Systematics* 33:427–447.
Kendal, Jeremy, Jamshid J. Tehrani, and John Odling-Smee
  2011 Human Niche Construction in Interdisciplinary Focus. *Philosophical Transactions of the Royal Society B* 366:785–792.
Kistler, Logan, Álvaro Montenegro, Bruce D. Smith, John A. Gifford, Richard E. Green, Lee A. Newsom, and Beth Shapiro
  2014 Transoceanic Drift and the Domestication of African Bottle Gourds in the Americas. *Proceedings of the National Academy of Sciences* 111:2937–2941.

Kistler, Logan, Lee A. Newsom, Timothy M. Ryan, Andrew C. Clarke, Bruce D. Smith, and George H. Perry
  2015 Gourds and Squashes (*Cucurbita* spp.) Adapted to Megafaunal Extinction and Ecological Anachronism through Domestication. *Proceedings of the National Academy of Sciences* 112:15107–15112.

Klein, Richard G.
  2013 Stable Carbon Isotopes and Human Evolution. *Proceedings of the National Academy of Sciences* 110:10470–10472.

Knight, Vernon James, Jr.
  1985 *Tukabatchee: Archaeological Investigations at an Historic Creek Town, Elmore County, Alabama.* Report of Investigations 45. University of Alabama Office of Archaeological Research, Alabama State Museum of Natural History, Tuscaloosa.

Kohler, Timothy A.
  2010 A New Paleoproductivity Reconstruction for Southwestern Colorado, and Its Implications for Understanding Thirteenth-Century Depopulation. In *Leaving Mesa Verde: Peril and Change in the Thirteenth-Century Southwest*, edited by Timothy A. Kohler, Mark D. Varien, and Aaron M. Wright, pp. 102–127. Amerind Studies in Archaeology 5. University of Arizona Press, Tucson.

Kohler, Timothy A., and Meredith H. Matthews
  1988 Long-Term Anasazi Land Use and Forest Reduction: A Case Study from Southwest Colorado. *American Antiquity* 53:537–564.

Kohler, Timothy A., R. Kyle Bocinsky, Denton Cockburn, Stefani A. Crabtree, Mark D. Varien, Kenneth E. Kolm, Schaun Smith, Scott G. Ortman, and Ziad Kobti
  2012 Modelling Prehispanic Pueblo Societies in Their Ecosystems. *Ecological Modelling* 241:30–41.

Kohler, Timothy A., Matt Pier Glaude, Jean-Pierre Bocquet-Appel, and Brian M. Kemp
  2008 The Neolithic Demographic Transition in the U.S. Southwest. *American Antiquity* 73:645–669.

Kohler, Timothy A., and Kelsey M. Reese
  2014 Long and Spatially Variable Neolithic Demographic Transition in the North American Southwest. *Proceedings of the National Academy of Sciences* 111:10101–10106.

Kohler, Timothy A., Mark D. Varien, Aaron M. Wright, and Kristin A. Kuckelman
  2008 Mesa Verde Migrations. *American Scientist* 96:146–153.

Kuckelman, Kristin A.
  2010 The Depopulation of Sand Canyon Pueblo, a Large Ancestral Pueblo Village in Southwestern Colorado. *American Antiquity* 75:497–525.

Laland, Kevin, Blake Matthews, and Marcus W. Feldman
  2016 An Introduction to Niche Construction Theory. *Evolutionary Ecology* 30:191–202.

Lapham, Heather A.
  2011 Animals in Southeastern Native American Subsistence Economies. In *The Subsistence Economies of Indigenous North American Societies: A Handbook*, edited by Bruce D. Smith, pp. 401–429. Smithsonian Institution Scholarly Press, Washington, DC.

Larsen, Clark Spencer
  1997 *Bioarchaeology: Interpreting Behavior from the Human Skeleton.* Cambridge University Press, Cambridge.
Lawson, John
  1967 [1709] *A New Voyage to Carolina.* University of North Carolina Press, Chapel Hill.
LeBlanc, Steven A.
  2008 The Case for an Early Farmer Migration into the Greater American Southwest. In *Archaeology without Borders: Contact, Commerce, and Change in the U.S. Southwest and Northwestern Mexico,* edited by Laurie D. Webster and Maxine E. McBrinn, pp. 107–142. University Press of Colorado, Boulder.
Le Page du Pratz, Antoine Simon
  1972 [1758] *The History of Louisiana or of the Western Parts of Virginia and Carolina.* English edition of 1774. Claitor's Publishing Division, Baton Rouge, Louisiana.
Lepofsky, Dana, Douglas Hallett, Ken Lertzman, Rolf Mathewes, Albert (Sonny) McHalsie, and Kevin Washbrook
  2005 Documenting Precontact Plant Management on the Northwest Coast: An Example of Prescribed Burning in the Central and Upper Fraser Valley, British Columbia. In *Keeping It Living: Traditions of Plant Use and Cultivation on the Northwest Coast of North America,* edited by Douglas Deur and Nancy J. Turner, pp. 218–239. University of Washington Press, Seattle.
Lepofsky, Dana, Nicole F. Smith, Nathan Cardinal, John Harper, Mary Morris, Gitla (Elroy White), Randy Bouchard, Dorothy I. D. Kennedy, Anne K. Salomon, Michelle Puckett, and Kirsten Rowell
  2015 Ancient Shellfish Mariculture on the Northwest Coast of North America. *American Antiquity* 80:236–259.
Lindgren, W. H., III
  1972 Agricultural Propaganda in Lawson's "A New Voyage to Carolina". *North Carolina Historical Review* 49:333–344.
Linton, Ralph
  1924 The Significance of Certain Traits in North American Maize Culture. *American Anthropologist* 26:345–359.
Lipe, William D., R. Kyle Bocinsky, Brian S. Chisholm, Robin Lyle, David M. Dove, R. G. Matson, Elizabeth Jarvis, Kathleen Judd, and Brian M. Kemp
  2017 Cultural and Genetic Contexts for Early Turkey Domestication in the Northern Southwest. *American Antiquity* 81:97–113.
Little, Elizabeth, and Margaret J. Schoeninger
  1995 The Late Woodland Diet on Nantucket Island and the Problem of Maize in Coastal New England. *American Antiquity* 60:351–368.
Lopinot, Neal H.
  1986 The Spanish Introduction of New Cultigens into the Greater Southwest. *Missouri Archaeologist* 47:61–84.
  1992 Spatial and Temporal Variability in Mississippian Subsistence: The Archaeobotanical Record. In *Late Prehistoric Agriculture: Observations from the Midwest,*

edited by William I. Woods, pp. 44–94. Studies in Illinois Archaeology 8. Illinois Historic Preservation Agency, Springfield.

1994 A New Crop of Data on the Cahokian Polity. In *Agricultural Origins and Development in the Midcontinent*, edited by William Green, pp. 127–154. Report 19. Office of the State Archaeologist, University of Iowa, Iowa City.

Mabry, Jonathan B.
2005 Changing Knowledge and Ideas about the First Farmers in Southeastern Arizona. In *The Late Archaic across the Borderlands: From Foraging to Farming*, edited by Bradley J. Vierra, pp. 41–83. University of Texas Press, Austin.

Mabry, Jonathan B., and William E. Doolittle
2008 Modeling the Early Agricultural Frontier in the Desert Borderlands. In *Archaeology without Borders: Contact, Commerce, and Change in the U.S. Southwest and Northwestern Mexico*, edited by Laurie D. Webster and Maxine E. McBrinn, pp. 55–70. University Press of Colorado, Boulder.

McLauchlan, Kendra
2003 Plant Cultivation and Forest Clearance by Prehistoric North Americans: Pollen Evidence from Fort Ancient, Ohio, USA. *The Holocene* 13:557–566.

Madsen, David B.
1998 The Fremont Complex. *Journal of World Prehistory* 12:255–336.

Marquardt, William H.
1974 Statistical Analysis of the Constituents in Human Paleofecal Specimens from Mammoth Cave. In *Archaeology of the Mammoth Cave Area*, edited by Patty Jo Watson, pp. 193–202. Academic Press, New York.

Marston, John M.
2011 Archaeological Markers of Agricultural Risk Management. *Journal of Anthropological Archaeology* 30:190–205.

Mathew, Sarah, and Charles Perreault
2015 Behavioural Variation in 172 Small-Scale Societies Indicates that Social Learning Is the Main Mode of Human Adaptation. *Proceedings of the Royal Society B-Biological Sciences* 282. Electronic document, http://rspb.royalsocietypublishing.org/content/282/1810/20150061, accessed February 13, 2017.

Matsuoka, Yoshihiro, Yves Vigouroux, Major M. Goodman, Jesus Sanchez G., Edward Buckler, and John Doebley
2002 A Single Domestication for Maize Shown by Multilocus Microsatellite Genotyping. *Proceedings of the National Academy of Sciences* 99:6080–6084.

Merrill, William L., Robert J. Hard, Jonathan B. Mabry, Gayle J. Fritz, Karen R. Adams, John R. Roney, and A. C. MacWilliams
2009 The Diffusion of Maize to the Southwestern United States and Its Impact. *Proceedings of the National Academy of Sciences A* 106:21019–21026.

Milner, George R.
1991 Health and Cultural Change in the Late Prehistoric American Bottom, Illinois. In *What Mean These Bones? Studies in Southeastern Bioarchaeology*, edited by Mary Lucas Powell, Patricia S. Bridges, and Ann Marie Wagner Mires, pp. 52–69. University of Alabama Press, Tuscaloosa.

Milner, George R., and Wirt H. Wills
  2013 Complex Societies of North America. In *The Human Past: World Prehistory & the Development of Human Societies*, edited by Chris Scarre, pp. 678–715. 3rd ed. Thames and Hudson, London.

Minnis, Paul E.
  1978 Paleoethnobotanical Indicators of Prehistoric Environmental Disturbance: A Case Study. In *The Nature and Status of Ethnobotany*, edited by Richard I. Ford, pp. 347–366. Anthropological Papers No. 67. Museum of Anthropology, University of Michigan, Ann Arbor.
  1985 *Social Adaptation to Food Stress: A Prehistoric Southwestern Example*. University of Chicago Press, Chicago.
  1989 Prehistoric Diet in the Northern Southwest: Macroplant Remains from Four Corners Feces. *American Antiquity* 54:543–563.
  1991 Famine Foods of the Northern American Desert Borderlands in Historical Context. *Journal of Ethnobiology* 11:231–257.
  2015 What More We Need to Know about "Southwestern" Agriculture. In *Traditional Arid Lands Agriculture: Understanding the Past for the Future*, edited by Scott E. Ingram and Robert C. Hunt, pp. 358–370. University of Arizona Press, Tucson.

Minnis, Paul E., Michael E. Whalen, and R. Emerson Howell
  2006 Fields of Power: Upland Farming in the Prehispanic Casas Grandes Polity, Chihuahua, Mexico. *American Antiquity* 71:707–722.

Morgan, Christopher, and Robert L. Bettinger
  2012 Great Basin Foraging Strategies. In *Oxford Handbook of North American Archaeology*, edited by Timothy R. Pauketat. Online ed. Oxford University Press, Oxford.

Mt. Pleasant, Jane
  2006 The Science behind the Three Sisters Mound System: An Agronomic Assessment of the Indigenous Agricultural System in the Northeast. In *Histories of Maize: Multidisciplinary Approaches to the Prehistory, Linguistics, Biogeography, Domestication, and Evolution of Maize*, edited by John E. Staller, Robert H. Tykot and Bruce F. Benz, pp. 529–537. Left Coast Press, Walnut Creek, California.

Mueller, Natalie G.
  2016a Documenting Domestication in a Lost Crop (*Polygonum erectum* L.): Evolutionary Bet-Hedgers under Cultivation. *Vegetation History and Archaeobotany* 26:313–327.
  2016b Evolutionary "Bet-Hedgers" under Cultivation: Investigating the Domestication of Erect Knotweed (*Polygonum erectum* L.) Using Growth Experiments. *Human Ecology* 45:189–203.
  2017 An Extinct Domesticated Subspecies of Erect Knotweed in Eastern North America: *Polygonum erectum* subsp. *watsoniae* (Polygonaceae). *Novon: A Journal for Botanical Nomenclature* 25:166–179.

Mueller, Natalie G., Gayle J. Fritz, Paul Patton, Stephen Carmody, and Elizabeth T. Horton
  2017 Growing the Lost Crops of Eastern North America's Original Agricultural System. *Nature Plants* 3:17092.

Munoz, Samuel E., Kristine E. Gruley, Ashtin Massie, David A. Fike, Sissel Schroeder, and John W. Williams
   2015 Cahokia's Emergence and Decline Coincided with Shifts of Flood Frequency on the Mississippi River. *Proceedings of the National Academy of Sciences* 112:6319–6324.

Munro, Natalie D.
   2011 Domestication of the Turkey in the American Southwest. In *The Subsistence Economies of Indigenous North American Societies: A Handbook*, edited by Bruce D. Smith, pp. 543–555. Smithsonian Institution Scholarly Press, Washington, DC.

Murray, Priscilla M., and Mark C. Sheehan
   1984 Prehistoric *Polygonum* Use in the Midwestern United States. In *Experiments and Observations on Aboriginal Wild Plant Food Utilization in Eastern North America*, edited by Patrick J. Munson, pp. 282–298. Prehistory Research Series Volume 6. Indiana Historical Society, Indianapolis.

Nabhan, Gary P., and J. M. J. deWet
   1984 *Panicum sonorum* in Sonoran Desert Agriculture. *Economic Botany* 38:65–82.

Nabhan, Gary P., Amadeo Rea, Karen L. Reichhardt, Eric Mellink, and Charles F. Hutchinson
   1982 Papago Influences on Habitat and Biotic Diversity: Quitovac Oasis Ethnoecology. *Journal of Ethnobiology* 2:124–143.

Nabhan, Gary P., Alfred Whiting, Henry Dobyns, Richard Hevly, and Robert Euler
   1981 Devil's Claw Domestication: Evidence from Southwestern Indian Fields. *Journal of Ethnobiology* 1:135–164.

Nelson, Margaret C.
   1999 *Mimbres during the Twelfth Century: Abandonment, Continuity, and Reorganization*. University of Arizona Press, Tucson.

Nelson, Margaret C., Michelle Hegmon, Keith W. Kintigh, Ann P. Kinzig, Ben A. Nelson, John Marty Anderies, David A. Abbott, Katherine A. Spielmann, Scott E. Ingram, Matthew A. Peeples, Stephanie Kulow, Colleen A. Strawhacker, and Cathryn Meegan
   2012 Long-Term Vulnerability and Resilience: Three Examples from Archaeological Study in the Southwestern United States and Northern Mexico. In *Surviving Sudden Environmental Change: Understanding Hazards, Mitigating Impacts, Avoiding Disasters*, edited by Jago Cooper and Payson Sheets, pp. 197–220. University Press of Colorado, Boulder.

Nelson, Margaret C., Keith Kintigh, David R. Abbott, and John M. Anderies
   2010 The Cross-Scale Interplay between Social and Biophysical Context and the Vulnerability of Irrigation-Dependent Societies: Archaeology's Long-Term Perspective. *Ecology and Society* 15:31.

Newbold, Bradley A., Joel C. Janetski, Mark L. Bodily, and David T. Yoder
   2012 Early Holocene Turkey (*Meleagris gallopavo*) Remains from Southern Utah. *Kiva* 78:37–60.

Newsom, Lee A., and D. Anne Trieu Gahr
   2011 Fusion Gardens: Native North America and the Columbian Exchange. In *The Subsistence Economies of Indigenous North American Societies: A Handbook*, edited by

Bruce D. Smith, pp. 557–576. Smithsonian Institution Scholarly Press, Washington, DC.

Odling-Smee, John, Kevin N. Laland, and Marcus W. Feldman
   1996   Niche Construction. *American Naturalist* 147:641–648.
   2003   *Niche Construction*. Monographs in Population Biology. Princeton University Press, Princeton, New Jersey.

Pauketat, Timothy R., Lucretia S. Kelly, Gayle J. Fritz, Neal H. Lopinot, Scott Elias, and Eve Hargrave
   2002   The Residues of Feasting and Public Ritual at Early Cahokia. *American Antiquity* 67:257–279.

Pavao-Zuckerman, Barnet, and Elizabeth J. Reitz
   2011   Eurasian Domesticated Livestock in Native American Economies. In *The Subsistence Economies of Indigenous North American Societies: A Handbook*, edited by Bruce D. Smith, pp. 577–591. Smithsonian Institution Scholarly Press, Washington, DC.

Payne, Willard, and Volney H. Jones
   1962   The Taxonomic Status and Archaeological Significance of a Giant Ragweed from Prehistoric Bluff Shelters in the Ozark Plateau Region. *Papers of the Michigan Academy of Science, Arts, and Letters* 47:147–163.

Peterson, James B., and Nancy Asch Sidell
   1996   Mid-Holocene Evidence of *Cucurbita* Sp. from Central Maine. *American Antiquity* 61:685–698.

Piperno, Dolores R.
   2016   Phytolith Radiocarbon Dating in Archaeological and Paleoecological Research: A Case Study of Phytoliths from Modern Neotropical Plants and a Review of the Previous Dating Evidence. *Journal of Archaeological Science* 68:54–61.

Piperno, Dolores R., and Kent V. Flannery
   2001   The Earliest Archaeological Maize (*Zea mays* L.) from Highland Mexico: New Accelerator Mass Spectrometry Dates and Their Implications. *Proceedings of the National Academy of Sciences* 98:2101–2103.

Popper, Virginia S.
   1988   Selecting Quantitative Measurements in Paleoethnobotany. In *Current Paleoethnobotany: Analytical Methods and Cultural Interpretations of Archaeological Plant Remains*, edited by Christine A. Hastorf and Virginia S. Popper, pp. 53–71. University of Chicago Press, Chicago.

Powell, Mary Lucas
   1991   Ranked Status and Health in the Mississippian Chiefdom at Moundville. In *What Mean These Bones? Studies in Southeastern Bioarchaeology*, edited by Mary Lucas Powell, Patricia S. Bridges, and Ann Marie Wagner Mires, pp. 22–51. University of Alabama Press, Tuscaloosa.

Rafferty, Sean M.
   2006 Evidence of Early Tobacco in Northeastern North America? *Journal of Archaeological Science* 33:453–458.

Ramenovsky, Ann F.
   1987 *Vectors of Death: The Archaeology of European Contact.* University of New Mexico Press, Albuquerque.

Reimer, Paula J., Edouard Bard, Alex Bayliss, J. Warren Beck, Paul G. Blackwell, Christopher Bronk Ramsey, Caitlin E. Buck, Hai Cheng, R. Lawrence Edwards, Michael Friedrich, Pieter M. Grootes, Thomas P. Guilderson, Haflidi Haflidason, Irka Hajdas, Christine Hatté, Timothy J. Heaton, Dirk L. Hoffmann, Alan G. Hogg, Konrad A. Hughen, K. Felix Kaiser, Bernd Kromer, Sturt W. Manning, Mu Niu, Ron W. Reimer, David A. Richards, E. Marian Scott, John R. Southon, Richard A. Staff, Christian S. M. Turney, and Johannes van der Plicht
   2013 IntCal13 and Marine13 Radiocarbon Age Calibration Curves 0–50,000 Years cal BP. *Radiocarbon* 55: 1869–1887.

Reinhard, Karl J., and Vaughn M. Bryant
   1992 Coprolite Analysis: A Biological Perspective on Archaeology. *Archaeological Method and Theory* 4:245–286.

Reitz, Elizabeth J.
   1991 Evidence for Animal Use at the Missions of Spanish Florida. *Florida Anthropologist* 44:299–306.

Reitz, Elizabeth J., and C. Margaret Scarry
   1985 *Reconstructing Historic Subsistence with an Example from Sixteenth-Century Spanish Florida.* Special Publication Series 3. Society for Historical Archaeology, Ann Arbor, Michigan.

Rhode, David
   2003 Coprolites from Hidden Cave, Revisited: Evidence for Site Occupation History, Diet and Sex of Occupants. *Journal of Archaeological Science* 30:909–922.

Richerson, Peter J., and Robert Boyd
   2005 *Not by Genes Alone: How Culture Transformed Human Evolution.* University of Chicago Press, Chicago.

Richerson, Peter J., Robert Boyd, and Joseph Henrich
   2010 Gene-Culture Coevolution in the Age of Genomics. *Proceedings of the National Academy of Sciences* 107:8985–8992.

Rindos, David
   1984 *The Origins of Agriculture: An Evolutionary Perspective.* Academic Press, New York.

Rindos, David, and Sissel Johannessen
   1991 Human-Plant Interactions and Cultural Change in the American Bottom. In *Cahokia and the Hinterlands: Middle Mississippian Cultures of the Midwest*, edited by Thomas E. Emerson and R. Barry Lewis, pp. 35–45. University of Illinois Press, Urbana.

Rocek, Thomas R.
   1995 Sedentarization and Agricultural Dependence: Perspectives from the Pithouse-to-Pueblo Transition in the American Southwest. *American Antiquity* 60:218–239.

Rogers, J. Daniel
  2011 Stable Isotope Analysis and Diet in Eastern Oklahoma. *Southeastern Archaeology* 30:96–107.
Rose, Fionnuala
  2008 Intra-community Variation in Diet during the Adoption of a New Staple Crop in the Eastern Woodlands. *American Antiquity* 73:413–439.
Rose, Jerome C., Murray K. Marks, and Larry L. Tieszen
  1991 Bioarchaeology and Subsistence in the Central and Lower Portions of the Mississippi Valley. In *What Mean These Bones? Studies in Southeastern Bioarchaeology*, edited by Mary Lucas Powell, Patricia S. Bridges, and Ann Marie Wagner Mires, pp. 7–21. University of Alabama Press, Tuscaloosa.
Roth, Barbara J.
  2016 *Agricultural Beginnings in the American Southwest*. Rowman & Littlefield, Lanham, Maryland.
Roth, Barbara J., and Andrea Freeman
  2008 The Middle Archaic Period and the Transition to Agriculture in the Sonoran Desert of Southern Arizona. *Kiva* 73:321–353.
Ruhl, Donna L.
  1993 Old Customs and Traditions in New Terrain: Sixteenth- and Seventeenth-Century Archaeobotanical Data from La Florida. In *Foraging and Farming in the Eastern Woodlands*, edited by C. Margaret Scarry, pp. 255–284. University Press of Florida, Gainesville.

Sahlins, Marshall, and Elman R. Service (editors)
  1960 *Evolution and Culture*. University of Michigan Press, Ann Arbor.
Sauer, Jonathan D.
  1994 *Historical Geography of Crop Plants: A Select Roster*. CRC Press, Boca Raton, Florida.
Saunders, Joe W., Rolfe D. Mandel, Roger T. Saucier, E. Thurman Allen, C. T. Hallmark, Jay K. Johnson, Edwin H. Jackson, Charles M. Allen, Gary L. Stringer, Douglas S. Frink, James K. Feathers, Stephen Williams, Kristen J. Gremillion, Malcolm F. Vidrine, and Reca Jones
  1997 A Mound Complex in Louisiana at 5400–5000 Years Before the Present. *Science* 277:1796–1799.
Scarry, C. Margaret
  1985 The Use of Plant Foods in Sixteenth Century St. Augustine. *Florida Anthropologist* 38:70–80.
  1991 Plant Production and Procurement in Apalachee Province. *Florida Anthropologist* 44:285–294.
  1993 Variability in Mississippian Crop Production Strategies. In *Foraging and Farming in the Eastern Woodlands*, edited by C. Margaret Scarry, pp. 78–90. University Press of Florida, Gainesville.
Scarry, C. Margaret, and Elizabeth J. Reitz
  1990 Herbs, Fish, Scum, and Vermin: Subsistence Strategies in Sixteenth-Century Spanish Florida. In *Columbian Consequences,* Volume 2: *Archaeological and Histori-*

cal Perspectives on the Spanish Borderlands East, edited by David Hurst Thomas, pp. 343–354. Smithsonian Institution Press, Washington, DC.

Scarry, C. Margaret, and John Scarry
    2005 Native American 'Garden Agriculture' in Southeastern North America. *World Archaeology* 37:259–274.

Scarry, C. Margaret, and Richard A. Yarnell
    2011 Native American Domestication and Husbandry of Plants in Eastern North America. In *The Subsistence Economies of Indigenous North American Societies: A Handbook*, edited by Bruce D. Smith, pp. 483–501. Smithsonian Institution Scholarly Press, Washington, DC.

Schachner, Gregson
    2001 Ritual Control and Transformation in Middle-Range Societies: An Example from the American Southwest. *Journal of Anthropological Archaeology* 20:168–194.

Schollmeyer, Karen Gust
    2005 Prehispanic Environmental Impact in the Mimbres Region, Southwestern New Mexico. *Kiva* 70:375–398.
    2011 Large Game, Agricultural Land, and Settlement Pattern Change in the Eastern Mimbres Area, Southwest New Mexico. *Journal of Anthropological Archaeology* 30:402–415.

Schollmeyer, Karen Gust, and Christy G. Turner
    2004 Dental Caries, Prehistoric Diet, and the Pithouse-to-Pueblo Transition in Southwestern Colorado. *American Antiquity* 69:569–582.

Schroeder, M., J. Deli, E. D. Schall, and G. G. Warren
    1974 Seed Composition of 66 Weed and Crop Species. *Weed Science* 22:345–348.

Schroeder, Sissel
    1999 Maize Productivity in the Eastern Woodlands and Great Plains of Eastern North America. *American Antiquity* 64:499–516.
    2001 Understanding Variation in Prehistoric Agricultural Productivity: The Importance of Distinguishing among Potential, Available, and Consumptive Yields. *American Antiquity* 66:517–525.

Service, Elman R.
    1975 *Origins of the State and Civilization: The Process of Cultural Evolution*. Norton, New York.

Sheldon, Elisabeth S.
    1978 Childersburg: Evidence of European Contact Demonstrated by Archaeological Plant Remains. *Southeastern Archaeological Conference Bulletin* 5:28–29.

Simms, Steven R.
    1987 *Behavioral Ecology and Hunter-Gatherer Foraging: An Example from the Great Basin*. International Series Volume 381. British Archaeological Reports, Oxford.

Simon, Mary L.
    2017 Reevaluating the Evidence for Middle Woodland Maize from the Holding Site. *American Antiquity* 82:140–150.

Simon, Mary L., and Kathryn E. Parker
    2006 Prehistoric Plant Use in the American Bottom: New Thoughts and Interpretations. *Southeastern Archaeology* 25:212–257.

Smith, Bruce D.
1985a *Chenopodium berlandieri* ssp. *jonesanum*: Evidence for a Hopewellian Domesticate from Ash Cave, Ohio. *Southeastern Archaeology* 4:107–133.
1985b The Role of *Chenopodium* as a Domesticate in the Pre-Maize Garden Systems of the Eastern United States. *Southeastern Archaeology* 4:51–72.
1987a The Economic Potential of *Chenopodium berlandieri* in Prehistoric Eastern North America. *Journal of Ethnobiology* 7:29–54.
1987b Independent Domestication of Indigenous Seed-Bearing Plants in Eastern North America. In *Emergent Horticultural Economies of the Eastern Woodlands*, edited by William F. Keegan, pp. 3–47. Occasional Paper 7. Center for Archaeological Investigations, Southern Illinois University, Carbondale.
1989 Origins of Agriculture in Eastern North America. *Science* 246:1566–1571.
1992a The Economic Potential of *Iva annua* in Eastern North America. In *Rivers of Change: Essays on Early Agriculture in Eastern North America*, edited by Bruce D. Smith, pp. 185–200. Smithsonian Institution Press, Washington, DC.
1992b The Floodplain Weed Theory of Plant Domestication in Eastern North America. In *Rivers of Change: Essays on Early Agriculture in Eastern North America*, edited by Bruce D. Smith, pp. 19–34. Smithsonian Institution Press, Washington, DC.
1992c Hopewellian Farmers of Eastern North America. In *Rivers of Change: Essays on Early Agriculture in Eastern North America*, edited by Bruce D. Smith, pp. 201–248. Smithsonian Institution Press, Washington, DC.
1992d In Search of *Choupichoul*, the Mystery Grain of the Natchez. In *Rivers of Change: Essays on Early Agriculture in Eastern North America*, edited by Bruce D. Smith, pp. 249–264. Smithsonian Institution Press, Washington, DC.
2001 Low-Level Food Production. *Journal of Archaeological Research* 9:1–43.
2005a Documenting the Transition to Food Production along the Borderlands. In *The Late Archaic across the Borderlands: From Foraging to Farming*, edited by Bradley J. Vierra, pp. 300–316. University of Texas Press, Austin.
2005b Low-Level Food Production on the Northwest Coast. In *Keeping It Living: Traditions of Plant Use and Cultivation on the Northwest Coast of North America*, edited by Douglas Deur and Nancy J. Turner, pp. 37–66. University of Washington Press, Seattle.
2006a Documenting Domesticated Plants in the Archaeological Record. In *Documenting Domestication: New Genetic and Archaeological Paradigms*, edited by Melinda A. Zeder, Daniel G. Bradley, Eve Emshwiller, and Bruce D. Smith, pp. 15–24. University of California Press, Berkeley.
2006b Eastern North America as an Independent Center of Plant Domestication. *Proceedings of the National Academy of Sciences* 103:12223–12228.
2009 Resource Resilience, Human Niche Construction, and the Long-Term Sustainability of Pre-Columbian Subsistence Economies in the Mississippi River Valley Corridor. *Journal of Ethnobiology* 29:167–183.
2011a The Cultural Context of Plant Domestication in Eastern North America. *Current Anthropology* 52 (Supplement 4):S471-S483.
2011b Introduction: Indigenous North American Societies and the Environment. In *The Subsistence Economies of Indigenous North American Societies: A Handbook*,

edited by Bruce D. Smith, pp. 1–9. Smithsonian Institution Scholarly Press, Washington, DC.

2011c General Patterns of Niche Construction and the Management of 'Wild' Plant and Animal Resources by Small-Scale Pre-Industrial Societies. *Philosophical Transactions of the Royal Society B* 366:836–848.

2014 The Domestication of *Helianthus annuus* L. (Sunflower). *Vegetation History and Archaeobotany* 23:57–74.

Smith, Bruce D., and C. Wesley Cowan

2003 Domesticated Crop Plants and the Evolution of Food Producing Economies in the Eastern United States. In *People and Plants in Ancient Eastern North America*, edited by Paul E. Minnis, pp. 105–125. Smithsonian Institution Press, Washington, DC.

Smith, Bruce D., C. Wesley Cowan, and Michael P. Hoffman

1992 Is It an Indigene or a Foreigner? In *Rivers of Change: Essays on Early Agriculture in Eastern North America*, edited by Bruce D. Smith, pp. 67–102. Smithsonian Institution Press, Washington, DC.

Smith, Bruce D., and Vicki A. Funk

1985 A Newly Described Subfossil Cultivar of *Chenopodium* (Chenopodiaceae). *Phytologia* 57:445–448.

Smith, Bruce D., and Richard A. Yarnell

2009 Initial Formation of an Indigenous Crop Complex in Eastern North America at 3800 B.P. *Proceedings of the National Academy of Sciences* 106:6561–6566.

Smith, Eric Alden

1979 Human Adaptation and Energetic Efficiency. *Human Ecology* 7:53–74.

Smith, Marvin

1987 *Archaeology of Aboriginal Culture Change in the Interior Southeast*. University of Florida Press, Gainesville.

Sobolik, Kristin D., Kristen J. Gremillion, Patricia L. Whitten, and Patty Jo Watson

1996 Technical Note: Sex Determination of Prehistoric Human Paleofeces. *American Journal of Physical Anthropology* 101:283–290.

Spielmann, Katherine A.

2008 Legacies on the Landscape: The Enduring Effects of Long-Term Human-Ecosystem Interactions. In *Movement, Connectivity, and Landscape Change in the Ancient Southwest: The 20th Anniversary Southwest Symposium*, edited by Margaret C. Nelson and Colleen A. Strawhacker, pp. 199–217. University Press of Colorado, Boulder.

Steward, Julian H.

1933 *Ethnography of the Owens Valley Paiute*. University of California Publications in American Archaeology and Ethnology 33. University of California Press, Berkeley.

Stewart, Robert B.

1974 Identification and Quantification of Components in Salts Cave Paleofeces, 1970–1972. In *Archaeology of the Mammoth Cave Area*, edited by Patty Jo Watson, pp. 41–48. Academic Press, New York.

Story, Dee Ann
   1985   Adaptive Strategies of Archaic Cultures of the West Gulf Coastal Plain. In *Prehistoric Food Production in North America*, edited by Richard I. Ford, pp. 19–56. Anthropological Papers No. 75. Museum of Anthropology, University of Michigan, Ann Arbor.
Struever, Stuart
   1968   Flotation Techniques for the Recovery of Small-Scale Archaeological Remains. *American Antiquity* 33:353–362.
Styles, Bonnie T.
   2011   Animal Use by Holocene Aboriginal Societies of the Northeast. In *The Subsistence Economies of Indigenous North American Societies: A Handbook*, edited by Bruce D. Smith, pp. 449–481. Smithsonian Institution Scholarly Press, Washington, DC.
Sullivan, Alan P.
   2015   The Archaeology of Ruderal Agriculture. In *Traditional Arid Lands Agriculture: Understanding the Past for the Future*, edited by Scott E. Ingram and Robert C. Hunt, pp. 273–305. University of Arizona Press, Tucson.
Sullivan, Alan P., and Kathleen M. Forste
   2014   Fire-Reliant Subsistence Economies and Anthropogenic Coniferous Ecosystems in the Pre-Columbian Northern American Southwest. *Vegetation History and Archaeobotany* 23:135–151.

Thurston, Tina L., and Christopher T. Fisher
   2007a   Intensification, Innovation, and Change: New Perspectives and Future Directions. In *Seeking a Richer Harvest: The Archaeology of Subsistence Intensification, Innovation, and Change*, edited by Tina L. Thurston and Christopher T. Fisher, pp. 249–259. Springer US, New York.
   2007b   Seeking a Richer Harvest: An Introduction to the Archaeology of Subsistence Intensification, Innovation, and Change. In *Seeking a Richer Harvest: The Archaeology of Subsistence Intensification, Innovation, and Change*, edited by Tina L. Thurston and Christopher T. Fisher, pp. 1–21. Springer US, New York.
Trigger, Bruce G., William R. Swagerty, and Wilcomb E. Washburn
   1996   Entertaining Strangers: North America in the Sixteenth Century. In *The Cambridge History of the Native Peoples of the Americas*, edited by Bruce G. Trigger and Wilcomb E. Washburn, pp. 325–398. Online ed. Cambridge University Press, Cambridge.
Tucker, Bram
   2006   A Future Discounting Explanation for the Persistence of a Mixed Foraging-Horticulture Strategy among the Mikea of Madagascar. In *Behavioral Ecology and the Transition to Agriculture*, edited by Douglas J. Kennett and Bruce Winterhalder, pp. 22–40. University of California Press, Berkeley.
Turgeon, Laurier
   1998   French Fishers, Fur Traders, and Amerindians during the Sixteenth Century: History and Archaeology. *The William and Mary Quarterly* 55:585–610.

Turner, Nancy, and Sandra Peacock
   2005 Solving the Perennial Paradox: Ethnobotanical Evidence for Plant Resource Management on the Northwest Coast. In *Keeping It Living: Traditions of Plant Use and Cultivation on the Northwest Coast of North America*, edited by Douglas Deur and Nancy J. Turner, pp. 101–150. University of Washington Press, Seattle.

Tykot, Robert
   2006 Isotope Analyses and the Histories of Maize. In *Histories of Maize: Multidisciplinary Approaches to the Prehistory, Linguistics, Biogeography, Domestication, and Evolution of Maize*, edited by John E. Staller, Robert H. Tykot, and Bruce F. Benz, pp. 131–142. Left Coast Press, Walnut Creek, California.

USDA
   2016 US Department of Agriculture, Agricultural Research Service, Nutrient Data Laboratory. USDA National Nutrient Database for Standard Reference, Legacy. Electronic document, https://www.ars.usda.gov/northeast-area/beltsville-md-bhnrc/beltsville-human-nutrition-research-center/nutrient-data-laboratory/docs/usda-national-nutrient-database-for-standard-reference/ accessed April 24, 2018.

Van Dyke, Ruth
   1999 The Chaco Connection: Evaluating Bonito-Style Architecture in Outlier Communities. *Journal of Anthropological Archaeology* 18:471–506.

VanDerwarker, Amber M.
   2010 Correspondence Analysis and Principal Components Analysis as Methods for Integrating Archaeological Plant and Animal Remains. In *Integrating Zooarchaeology and Paleoethnobotany: A Consideration of Issues, Methods, and Cases*, edited by Amber M. VanDerwarker and Tanya M. Peres, pp. 75–95. Springer, New York.

Vierra, Bradley J., and Richard I. Ford
   2007 Foragers and Farmers in the Northern Rio Grande Valley, New Mexico. *Kiva* 73:117–130.

Vivian, R. Gwinn
   1990 *The Chacoan Prehistory of the San Juan Basin*. Academic Press, New York.

Wagner, Gail
   1986 The Corn and Cultivated Beans of the Fort Ancient Indians. *Missouri Archaeologist* 47:107–136.
   1987 Uses of Plants by the Fort Ancient Indians. PhD dissertation, Department of Anthropology, Washington University, St. Louis.
   2000 Tobacco in Prehistoric Eastern North America. In *Tobacco Use by Native North Americans: Sacred Smoke and Silent Killer*, edited by Joseph C. Winter, pp. 185–201. University of Oklahoma Press, Norman.
   2003 Eastern Woodlands Anthropogenic Ecology. In *People and Plants in Ancient Eastern North America*, edited by Paul E. Minnis, pp. 126–171. Smithsonian Institution Press, Washington, DC.

Wagner, Gail, and Peter H. Carrington
 2014 Sumpweed or Marshelder (*Iva annua*). In *New Lives for Ancient and Extinct Crops*, edited by Paul E. Minnis, pp. 65–101. University of Arizona Press, Tucson.

Waselkov, Gregory A.
 1989 Seventeenth-Century Trade in the Colonial Southeast. *Southeastern Archaeology* 8:117–133.

Watson, Patty Jo (editor)
 1974 *Archaeology of the Mammoth Cave Area*. Academic Press, New York.
 1985 The Impact of Early Horticulture in the Upland Drainages of the Midwest and Midsouth. In *Prehistoric Food Production in North America*, edited by Richard I. Ford, pp. 99–148. Anthropological Papers No. 75. Museum of Anthropology, University of Michigan, Ann Arbor.

Watson, Patty Jo, and Richard A. Yarnell
 1966 Archaeological and Paleoethnobotanical Investigations in Salts Cave, Mammoth Cave National Park, Kentucky. *American Antiquity* 31:842–849.
 1986 Lost John's Last Meal. *Missouri Archaeologist* 47:241–255.

Weiser, Andrea, and Dana Lepofsky
 2009 Ancient Land Use and Management of Ebey's Prairie, Whidbey Island, Washington. *Journal of Ethnobiology* 29:184–212.

Weitzel, Eric M., and Brian F. Codding
 2016 Population Growth as a Driver of Initial Domestication in Eastern North America. *Royal Society Open Science* 3:160319.

White, Leslie A.
 1959 *The Evolution of Culture: The Development of Civilization to the Fall of Rome*. McGraw-Hill, New York.

White, Nancy Marie, and Richard A. Weinstein
 2008 The Mexican Connection and the Far West of the U.S. Southeast. *American Antiquity* 73:227–277.

White, Richard
 1983 *The Roots of Dependency: Subsistence, Environment, and Social Change among the Choctaws, Pawnees, and Navajos*. University of Nebraska Press, Lincoln.

Wills, W. H.
 1988 Early Agriculture and Sedentism in the American Southwest: Evidence and Interpretations. *Journal of World Prehistory* 2:445–488.
 1995 Archaic Foraging and the Beginning of Food Production in the North American Southwest. In *Last Hunters, First Farmers: New Perspectives on the Prehistoric Transition to Agriculture*, edited by T. Douglas Price and Anne Birgitte Gebauer, pp. 215–242. School of American Research Press, Santa Fe, New Mexico.

Wilshusen, Richard H., and Scott G. Ortman
 1999 Rethinking the Pueblo I Period in the San Juan Drainage: Aggregation, Migration, and Cultural Diversity. *Kiva* 64:369–399.

Wilson, Diane, and Timothy K. Perttula
 2013 Reconstructing the Paleodiet of the Caddo through Stable Isotopes. *American Antiquity* 78:702–723.

Windingstad, Jason
  2006 Soil Quality and Indigenous Foodplot Locations in Eastern Kentucky. Master's thesis, Department of Geology, University of Tennessee, Knoxville.
Windingstad, J. D., S. C. Sherwood, K. J. Gremillion, and N. S. Eash
  2008 Soil Fertility and Slope Processes in the Western Cumberland Escarpment of Kentucky: Influences on the Development of Horticulture in the Eastern Woodlands. *Journal of Archaeological Science* 35:1717–1731.
Winterhalder, Bruce
  2012 Risk and Decision-Making. In *Oxford Handbook of Evolutionary Psychology*, edited by R. I. M. Dunbar and Louise Barrett. Oxford Handbooks Online, Oxford.
Winterhalder, Bruce, and Carol Goland
  1993 On Population, Foraging Efficiency, and Plant Domestication. *Current Anthropology* 34:710–715.
  1997 An Evolutionary Ecology Perspective on Diet Choice, Risk, and Plant Domestication. In *People, Plants, and Landscapes: Studies in Paleoethnobotany*, edited by Kristen J. Gremillion, pp. 123–160. University of Alabama Press, Tuscaloosa.
Winterhalder, Bruce, Flora Lu, and Bram Tucker
  1999 Risk-Sensitive Adaptive Tactics: Models and Evidence from Subsistence Studies in Biology and Anthropology. *Journal of Archaeological Research* 7:301–348.
Winterhalder, Bruce, and Eric Alden Smith
  1992 Evolutionary Ecology and the Social Sciences. In *Evolutionary Ecology and Human Behavior*, edited by Eric Alden Smith and Bruce Winterhalder, pp. 3–24. Aldine de Gruyter, New York.
  2000 Analyzing Adaptive Strategies: Human Behavioral Ecology at Twenty-Five. *Evolutionary Anthropology* 9:51–72.
Woodburn, James
  1982 Egalitarian Societies. *Man* 17:431–451.
Woods, William I.
  2004 Population Nucleation, Intensive Agriculture, and Environmental Degradation: The Cahokia Example. *Agriculture and Human Values* 21:255–261.
Wrangham, Richard W., James Holland Jones, Greg Laden, David Pilbeam, and Nancy-Lou Conklin-Brittain
  1999 The Raw and the Stolen: Cooking and the Ecology of Human Origins. *Current Anthropology* 40:567–594.
Wymer, Dee Anne
  1993 Cultural Change and Subsistence: The Middle Woodland and Late Woodland Transition in the Mid-Ohio Valley. In *Foraging and Farming in the Eastern Woodlands*, edited by C. Margaret Scarry, pp. 138–156. University Press of Florida, Gainesville.

Yarnell, Richard A.
  1964 *Aboriginal Relationships between Culture and Plant Life in the Upper Great Lakes Region*. Anthropological Papers No. 23. University of Michigan Museum of Anthropology, Ann Arbor.

1971 Early Woodland Plant Remains and the Question of Cultivation. In *Prehistoric Agriculture*, edited by Stuart Struever, pp. 550–554. Natural History Press, Garden City, New York.

1972 *Iva annua* var. *macrocarpa*: Extinct American Cultigen? *American Anthropologist* 74:335–341.

1974 Plant Foods and Cultivation of the Salts Cavers. In *Archaeology of the Mammoth Cave Area*, edited by Patty Jo Watson, pp. 113–122. Academic Press, New York.

1978 Domestication of Sunflower and Sumpweed in Eastern North America. In *The Nature and Status of Ethnobotany*, edited by Richard I. Ford, pp. 289–300. Anthropological Papers 67. Museum of Anthropology, University of Michigan, Ann Arbor.

1986 A Survey of Prehistoric Crop Plants in Eastern North America. *Missouri Archaeologist* 47:47–59.

Yarnell, Richard A., and M. Jean Black

1985 Temporal Trends Indicated by a Survey of Archaic and Woodland Plant Food Remains from Southeastern North America. *Southeastern Archaeology* 4:93–106.

Yerkes, Richard W.

2005 Bone Chemistry, Body Parts, and Growth Marks: Evaluating Ohio Hopewell and Cahokia Mississippian Seasonality, Subsistence, Ritual, and Feasting. *American Antiquity* 70:241–265.

Zeanah, David W.

2004 Sexual Division of Labor and Central Place Foraging: A Model for the Carson Desert of Western Nevada. *Journal of Anthropological Archaeology* 23:1–32.

Zeanah, David W., and Steven R. Simms

1999 Modeling the Gastric: Great Basin Subsistence Studies since 1982 and the Evolution of General Theory. In *Models for the Millennium: Great Basin Anthropology Today*, edited by Charlotte Beck, pp. 118–140. University of Utah Press, Salt Lake City.

Zeder, Melinda A.

2015 Core Questions in Domestication Research. *Proceedings of the National Academy of Sciences* 112:3191–3198.

Zeder, Melinda A., Eve Emshwiller, Bruce D. Smith, and Daniel G. Bradley

2006 Documenting Domestication: The Intersection of Genetics and Archaeology. *Trends in Genetics* 22:139–155.

Index

accelerator mass spectrometry (AMS), 25, 67
acculturation, 106, 120, 126, 133
acorn economies, 10
Adair, James, 65
adaptive syndrome of domestication, 25–26
Africans, enslaved, 113
agave, 46, 61
agriculture. *See also* maize agriculture; maize agriculture, in Eastern Woodlands; maize agriculture, in Southwest; pre-maize agriculture; *specific practices*
    definition of farming and, 2–3
    ecoregion of drylands, 9–10
    maize-based systems of, 78–84
    Native conversion to, 120
    Neolithic Demographic Transition and, 62–63
    northern Southwest settlement and, 56
    shifting cultivation versus permanent fields, 84–85
akchin fields, 55
amaranth, 22, 47, 48, 81
American Antiquity, 24
American Bottom, 13, 14, 32, *68*, 74, 80–81, 86
    deforestation and, 87
AMS. *See* accelerator mass spectrometry
Ancestral Puebloans, *39*, 55–59, 102
Andean potato, *112*, 113, *117*
Anderson, Edgar, 26

anemia, 75
animals, 53. *See also* livestock
    domestic, vii, viii, 129, 134
    husbandry, viii, 92, 107, 111, 118, 119
    population growth impact on, 60
annuals
    date of reliance on Great Plain, 8
    domestication of, 2
    Early Agricultural Southwest, 47
    weedy, 12, 15
Appalachian bean, 115
apatite, 74
Appalachian region, 6, *6*, 32, 33, 36, 114, 119, 142
Archaic period, 13, *14*, 88, 132
    EAC dispersal and development during late, 27–28
autonomy, Native, 121, 126
avoidance of risk, 46

ball courts, 62
barley, 22
barley grass, 61
Basketmaker area, 39, *40*, 43, 50
    bone chemistry in, 52
beans, 61, 78, 80, 115
    radiocarbon dates for maize, squash and, *44*
biogeography, 3
black-eyed pea. *See* cowpea
black walnut, 7
Bluff Dweller Culture, 23
Boas, Franz, 92

bone chemistry, Basketmaker area, 52
bone collagen, 74
botany. *See* plants; *specific regions*
broadcast sowing, 34, 65, 127
burning, controlled, 8, 48, 56, 86–87, 97, 102
    postcontact, 110
    in prehistory, 45

cacique fields, 60–61
cactus, 46, 47
Caddo area, 72
Cahokia site, *68*, 74, *81*
California. *See* Mediterranean California
camas, *96*, 97
canal irrigation, 45, 53, 55, 61, 62, 63, 78, 85
carbon isotopes
    in Eastern Woodlands, 68, 71
    maize consumption and, 51–53, *52*
    variation in nitrogen and, *52*
Casas Grandes (Paquimé), 55, 60–61
cave food, paleofecal indicators of, 30, *31*, 37*n*2
Cerro Juanaqueña, 51
Chacoan system, 56–57
chenopod, 36, 81
Chenopodium berlandieri. *See* goosefoot
chickens, 117
chocolate lily, 96
chroniclers, maize and, 65
clam gardens, 98
Classic Mimbres sites, 55, 59–60
Clearwater, *39*
    radiocarbon dates on maize from, *44*
Cliff Houses, 23
Cloudsplitter site, *20*
Coastal Plain, dominant plants of, 6
coevolution, 1, 3, 143
coevolutionary continuum, 1, 3, 143
    ecological context of, 4–11, *7*
colonization, 108–109. *See also* Spanish missions
Colorado Plateau, 38, 45, 50, 55, 56, 63, 100, 131

maize agriculture chronology in, *40*
    radiocarbon dates on maize from, *44*
Columbia Plateaus, 92, *94*, 100, 143
Columbus, Christopher, 108
Commission for Environmental Cooperation, *6*
conquistadores, 65
coevolutionary continuum, vii, 3
    ecological context of, 4–11, *7*
coprolites. *See* paleofeces
cowpea (black-eyed pea), 113, 115, 116
cribra orbitalia (porotic hyperostosis), 75
crops. *See also* Old World crops; plants; *specific crops*; *specific regions*; *specific topics*
    environmental constraints on, 4–5
    introduction of specialty, 117
    maize as, 42
Cucurbita pepo, 21, 24–25
cucurbits (gourds, squashes, melons), 116. *See also* squash
cultivated plants, 3
Cumberland Plateau, 28

deer, 53, 57, 58, 97
    whitetail, *7*, 34–35, 86
deforestation, 32, 87
demographic expansion, Southwest maize agriculture and, 62–63
demographic transitions
    Neolithic, 62–63
dental caries, 75
depopulation, Native, 57, 60, 63, 84
    postcontact, 106, 110–111
deserts, North American, *6*
    ecoregion features of, 9–10
de Soto, Hernando, 65, 110, 118
diet breadth model, 31, 128, 137
diets, Native
    challenge of inferring maize in, 51–52
    direct assessment of, 29–30
    Eastern Agricultural Complex and, 29–31, 37*n*2
    Europeanized, 120
    famine foods and, 46

human remains as indicator of, 71
maize in Early Agricultural, 64
postcontact, 115, 120
seasonal patterns and, 30
transition to maize-based, 71–76
direct dating, Eastern Woodland sites, 67–68
domesticates and domestication
in agriculture and farming terminology, 2
animal, vii, viii, 129, 134
of annuals and perennials, 2
chronological date of, 13
on coevolution continuum, 3
Eastern Agricultural Complex, *18*, 34, 129
FWT and adaptive syndrome of, 25–26
gold standard of evidence for, 67
initial adoption model of, 2, 125–126, 131–132
initial domestication of local species, 125, 126–131
natural selection and, ix
Old World, 107–108
pre-maize, 12–13, 37*n*1
radiocarbon dates for EAC, *20*
sites with earliest records of plant, 34
Doolittle, William, 84
dormancy trait, 26
drylands agriculture, ecoregion of, 9–10
dump heap theory, 26

EAC. *See* Eastern Agricultural Complex
Early Agricultural period, 46, 60–63
Ancestral Puebloans, *39*, 55–59, 102
farmer impact on land during, 53
initial adoption of domesticates and, 131–132
maize dietary reliance in, 64
maize intensification in, 138–140
native plant husbandry, 47–49
Southwest farming and, 45, 53
storage, 49, 50, 51
subsistence role of maize in, 49–53
Early Holocene period, *14*

Early Mesoamerican Crop Complex, 41
Early Pithouse period, 59
Early Southwestern Crop Complex, 41
East
initial adoption of maize in, 132
livestock introduction in, 118–119
maize intensification in, 140–142
temperate forests ecological region, 5–8, *6*
Eastern Agricultural Complex (EAC), vii, 11, 125
abandoned crops of, 80
adaptive syndrome of domestication theory and, 25–26
anthropogenic landscape effects and, 31–34
Archaic period dispersal and development, 27–28
chronology and culture history, 13–14, *14*
earliest radiocarbon dates for domesticated crops of, *20*
Eastern Woodlands secondary crops from, 65–66
farming communities, land tenure and, 34–35
farming technology and, 32–34
floodplain weed theory and, 25–26, 127
on food production continuum, 90
food production terminology and, 12–13
human diets and, 29–31
initial domestication of local species in, 126–130
key archaeological sites and river valleys in, *14*, *17*
maize introduction into, 37
maize yields compared with, 82–83
native seed crop intensification, 136–138
in North American prehistory, 22–25
nutrient composition of selected, *16*
plant life and ecology of crops in, 15–22, *18*, *19*, *116*

profitability and, 128, 129
regional variability in role of, 35–36
social complexity and, 34–35
subregion correlations, 13
topics under debate, 25
Woodland period food production in, 28–36
Eastern Woodlands. *See also* maize agriculture, in Eastern Woodlands
carbon isotope studies for, 68, 71
culture areas and chronology, *14*, 66–67
initial adoption model and, 125
secondary crops in, 65–66
site map for, *68*
Three Sisters crop complex and, 78, 80, 84
ecological regions, of North America, 5–11, *6*. *See also specific regions*
ecology
coevoluntionary continuum and, 4–11, *7*
EAC botany and, 15–22, *18*, *19*, *116*
evolutionary, 122–125
of maize-based farming, 82–87
Southwest subsistence, 4–11, *7*
traditional ecological knowledge, 123
weedy crop definition and, 3
erect knotweed, 15, *19*, 21–22
European contact, viii, 82, 105–107
European exploration and colonization, 108–109, *109*
first centuries of, 121
maize consumption predating, 76
postcontact landscape, 110–111
evolutionary ecology, 122–125

famine foods, 46
farming. *See also* agriculture; food production
continuum of foraging and, vii, 90, 91
definition of agriculture and, 2–3
dichotomy between foraging and, 1, 91, 92
diffuse origins of, 43
dry, 45

European, 126
residential permanence and, 54
upland, 32–34
Woodland period communities of, 34–35
feces. *See* paleofeces
fire. *See* burning
fish, 9, 66, 81, 93, 94, 97, 108, 127
flooding, 7–8, 33, 53, 61, 86, 87
floodplain weed theory (FWT), 25–26, 127
Florida, southern tip of, 6–7
flotation revolution, 24
food procurement
agriculture as opposite of, 2
pre-maize, 12–13
food production. *See also* agriculture; maize agriculture; *specific periods*; *specific regions*; *specific sites*
author of proposed pre-maize, 23
continuum, 90
definition and terminology, 2
diverse forms of, vii
Eastern Agricultural Complex term for, 12–13
important archaeological sites of, 14, *17*
nonagricultural, 90–92, 104, 126, 142–144
patterns of reliance on, x
postcontact, 105–108, 111–117, *112*
transition types in development of, 125–126
unrecognized, 9
foraging. *See also* Pacific Northwest
continuum of farming and, vii, 90, 91
dichotomy between farming and, 1, 91, 92
maize and, 131
optimal foraging theory, 31
reduction of, 11
social stratification and, 91
Southwest patterns of, 47
Fort Ancient area, 33, 75, 80–81, 88–89
Fremont area, 102–103, 139, 142

fruit, heteromorphism in, 22
FWT. *See* floodplain weed theory

gardens
    hunting in, 85
    second, 61
geophytes, 95–96
goosefoot (lambsquarters), 15, *16*, 17–19, *18*, 30, 47, 70, 81
    earliest evidence of weedy, 36
    floodplain weed theory and, 127
    maize yields compared to, 83
    radiocarbon dates for EAC, *20*
    at Riverton, 27
grasses
    of Eastern Agricultural Complex, 22
    maize relatives, 41
Great Basin, 91, *94*, 99, *143*
    Fremont culture and maize in, 102–103
    as human habitat, 100
    perennials, 101–102
    small seed complex in, 100–101
Greater Southwest
    condition summary prior to 2000 BC, 64
    drylands agriculture, 9–10
    livestock in, 117
    maize in, *39*, 131–132
    Old World crops introduction in, *112*, 116–117
    radiocarbon dates on maize, beans and squash from, *44*
Great Houses, 56, 57
Great Plains, 7
    ecoregion of, *6*, *8*
    livestock introduced in, 119
groundnut, 80

Harness, *68*
Harriot, Thomas, 65
Hart, John, 77
Haury, Emil, 49
Hayes site, 19, *20*
Haystack shelter, 30

hazelnut, 7
health
    maize impact on human, 75–76
    Old World pathogens and, 110
heteromorphism, in fruit, 22
hickory nuts, *16*, 27, 35, 127, 129, 142
hoe, Mill Creek chert, 86
Hohokam area, 39, *39*, 46, 54, 139
    fire use, 48
    irrigation, 45, 61–62
    maize agriculture and, 61–62
    plants farmed in, 61
Holding, 67, 68, *68*
Holocene, *14*, *143*
hominy, 77, 141
Hooton Hollow, 30
Hopewell era, 33, 35, 132
    definition of, 13
human diet. *See* diets, human
human remains, diet evidence in plants compared to, 71
hunter-gatherers, 90–91, 92
husbandry, Early Agricultural period native plant, 47–49. *See also* animals

Iberian crops, 111, 113, 120
Icehouse Bottom, *68*
imports, food, 57
initial adoption model, 2, 125–126, 131–132
intensification, vii, 10, 134–136
    in agriculture and farming terminology, 2
    contrasting rapid with gradual, 127
    date of maize most rapid, 66
    debate over, 144
    definition of, 2, 134–135
    delayed maize, 76–78
    EAC native seed crop, 136–138
    maize, Eastern, 140–142
    nonagricultural, 90–92, 104, 126, 142–144
    population growth and, 54
    Southwest maize agriculture, 138–140
intercropping, 80

iron-deficiency anemia, 75
irrigation
　canals, 45, 53, 55, 61, 62, 63, 78, 85
　child and infant mortality link with, 62
　Hohokam, 45, 61–62
　maize, 43, 45

Jones, Volney, 19, 23

knotweed, 15, *19*, 21–22, *36*

lambsquarters. *See* goosefoot
land tenure, 29, 34–35, 98
Langford Tradition, 74
Las Capas site, 50–51
Late Archaic period, 27–28
Late Holocene, *14*
Late Pleistocene period, *14*
Late Southwestern Complex, 48
Late Woodland period, 13
Linton, Ralph, 23
livestock, 107
　adoption of Old World crops and, 132–134
　eastern introduction of, 118–119
　regions where successful, 117–118
low-level food production
　examples of, 1–2
　of hunter-gatherers, 90–91, 92

maize. *See also specific regions*
　arrival in Southwest, vii
　closest relative, 41
　consumption evidence challenge, 51–53, *52*
　as crop, 42
　date of centrality, 14
　EAC introduction of, 37
　Eastern initial adoption of, 132
　environmental constraints on, 4–5
　in Great Basin Fremont area, 102–103
　human health and, 75–76
　importation, 57
　macroremains in Eastern North America, *67*
　maize early dates, 43
　Native diet and, 51–52, 71–76
　nixtamalization of, 70, 77, 141
　nutrient composition of EAC crops compared with, *16*
　nutrition-improving treatment of, 70
　origins of, vii, 4, 41–42
　in postcontact landscape, 121
　pre-maize agriculture, 12, 22–25, 37*n*1
　radiocarbon dates for beans, squash and, *44*
　requirements of modern, 42
　social status and, 75–76
　as turkey fodder, 59
　yields, 77, 82–83
maize agriculture
　diversity, viii
　ecoregions of, 9
　environmental constraints and, 4–5
　in North American deserts, 9–10
　onset of, vii
　pollen indicators of, 49
maize agriculture, in Eastern Woodlands, *68*
　agricultural systems and, 78–87
　Caddo area, 72
　chroniclers and, 65
　culture areas and chronology, *14*, 66–67
　delayed intensification explanations, 76–78
　differential preservation effects and, 77
　direct dating and, 67–68
　EAC crops and, 65
　earliest dates from macroremains in, *67*
　ecological impacts of maize-based farming, 82–87
　environmental impacts, 86–87
　European contact and, 76
　in-migration theory, 89
　land use and cropping systems, 83–84
　lesser crops and, 80–82
　Middle Ohio Valley, 72–73
　Mississippian polities and, 87–89

190

multiple introductions theory for, 77–78
Northeast area, 73
routes and means of introduction, 68–71
shifting cultivation versus permanent fields, 84–85
Southeast area of, 73
technological innovation and, 77
topographic modifications and, 85–86
transition to maize-based diets, 71–76
variation within communities, 74–75
yield improvement and, 77
maize agriculture, in Southwest AD 200-1400, 54–63
Ancestral Puebloans and, 55–59
anthropogenic impacts and, 53
bone chemistry and, 52
carbon and nitrogen isotope variation, 52
Chacoan system and, 56–57
chronology, *40*
culture areas and chronology, 38–41, *39*
demographic expansion and, 62–63
Early Agricultural period communities, 46–64
foraging patterns, 47
Hohokam area, 61–62
intensification, 138–140
irrigation, 43, 45
maize, other crop dispersal in, 41–44, *44*
Mogollon area, 59–61
residential permanency and, 54
strategies and risk management, 45–46
subsistence role of maize, 49–53
Mammoth Cave system, 23, 29
Marine West Coast Forests, 6, *8–9*
marshelder (sumpweed), 15, *16*, *18*, *20*, 128
masting, 130
maygrass, 22, 28, 70, 81
McEuen Cave
radiocarbon dates on maize from, *44*

Mediterranean California, 6, *10*
megadroughts, 63
melden, 65
microremains, plant, 69
middle ground, definition of, 2
Middle Ohio Valley, 72–73
Middle Woodland period, 13
Mid-Holocene period, *14*
floodplain weed theory and, 26
Mill Creek chert hoe, blade trading, 86
Mississippian polities, 87–89
Mississippi Valley, 73
Mogollon area, 38–39, *39*, 49
Classic Mimbres sites of, 55, 59–60
Paquimé, 60–61
radiocarbon dates on maize, beans and squash from, *44*
monocropping, 84
monument construction, 35
Mounded Talus Rockshelter, *18*
mounds, 80, 88
Moundville, *68*, 73, 141
Museum of Anthropology, University of Michigan, 23
mutual dependency (mutualism), ix–x
coevolution arising from, 1

Napoleon Hollow (West-central Illinois), *20*
native seed crops, 4
natural selection, domestication and, ix
Neolithic Demographic Transition (NDT), 62–63
Newt Kash, *20*, 23, 30
niche construction, 124, 129
nixtamalization, of maize, 70, 77, 141
North America, viii. *See also* food production, postcontact; *specific regions*
Eastern Agricultural Complex in prehistorical, 22–25
ecoregions, 5–11, *6*
maize introduction and dispersal in, *41*, 41–44
maize macroremains in Eastern, *67*

major culture areas and physiographic features, 7
postcontact food production in, 105–107
Northeast area
 livestock in, 119
 maize consumption in Eastern Woodlands, 73
Northern Flint, 140, 141

oaks, 6, 7
Old Corn site, *44*, *50*
Old World
 diseases, 110
 pathogens, 113
Old World crops, 106
 adoption of livestock and, 132–134
 domesticates, 107–108
 introduction of, 111–117, *112*
optimal foraging theory, 31

Pacific Coast, Mediterranean climate and ecoregion, *6*, 10
Pacific Northwest (Cascadia), 10, *94*, 126
 as anomalous, 93
 clam gardens of, 98
 cultural evolutionism and, 92–93
 as human habitat, 93–95
 land tenure, 98
 maritime management, 97–98
 plant management, 95–97
paleofeces (coprolites), 101
 diet inferred from, 29, 30, *31*, 37*n*2
Palmer Drought Severity Index (PDSI), 63
Paquimé. *See* Casas Grandes
Patayan region, *39*
PDSI. *See* Palmer Drought Severity Index
peaches, 113–114
peach pits, 114
Pecos Classification, 39, *40*
Pee Dee Belemnite, 72
Pennsylvania, 36
perennials
 in acorn economies, 10
 domestication and, 2

geophytes, 95–96
Great Basin, 101–102
permanent fields, shifting cultivation versus, 84–85
Phillips Spring (Missouri), *20*
phytoliths (silica bodies), 69
pines, codominance of, 6
piñon, 47, 49, 102
Pithouse-to-Pueblo transition, 54, 55–56, 59–60
Plains, 69
 postcontact, 116–117
Plains Village tradition, 8, 66
plants. *See also* crops; domesticates and domestication; *specific regions*
 defining wild and cultivated, 3
 diet evidence from human remains compared to, 71
 ecoregions of, 5–11, *6*
 microremains, 69
population growth, 55, 58, 63, 136–138. *See also* depopulation, Native
 animal impact from, 60
 intensification and, 54
 resource depression from, 31, 140
porotic hyperostosis. *See* cribra orbitalia
potatoes, *112*, 113, 115, 116, 117
pottery, as date marker, 13
pre-maize agriculture, 22–25, 37*n*1
 weedy annuals and, 12
pseudocereals, 15, 81
Pueblo Bonito, Great House, 57
Pueblo sequence, maize and, 39–40, *40*
 Hohokam area, 54
pumpkin. *See* squash

quinoa, 15

ragweed, 22, 32
rainfall, maize requirement for, 42
resource depression, 31, 140, 143
ridged fields, 86
risk, 45
 assessment, 123–124
 avoidance, 46

Riverton site, 27, 34
river valleys, EAC key, 14, *17*
Roundtop, *68*

salmon, 9, 93
second gardens, 61
semicultivation, 48
shifting cultivation, permanent fields versus, 84–85
silica bodies. *See* phytoliths
sites, with earliest records of plant domestication, 34. *See also specific sites*
slavery, African, 113
Smith, Bruce, 25–26
social learning, 123–125, 133–134
social stratification, 75–76, 87–88
   foraging and, 91
soil mounds system, 80, 88
Southeast
   livestock in, 118
   lower Mississippi Valley and, 73
   Old World crops in, 113–114
southeastern Coastal Plain, *6*, *7*
Southwest. *See also* Greater Southwest; maize agriculture, in Southwest
   agriculture and settlement in northern, 56
   defined area of, 38
   diverse crops of, 41
   maize arrival in, vii
   Neolithic Demographic Transition and, 62–63
   subsistence ecology in, 38
Southwestern Crop Complex, Early, 41
Spanish chroniclers, 65
Spanish missions, 108, 111, 126, 133
   crops, 116
   system, 119–121
squash, *18*, 19, *20*, 21
   definition of, 37n1
   maize remains accompanied by, 43
   pre-maize domestication of, 12, 37n1
   radiocarbon dates for maize, beans and, *44*

Three Sisters and, 78, 80, 84
Steward, Julian, 99–100
stone masonry, 56
storage, 28, 30, 31, 34, 130, 136, 138
   Early Agricultural period, 51
   pits, 49, 50
succulents, 47
sumpweed. *See* marshelder
sunflower, 36, 65, 128
   domestic, *15*, *18*, *19*, *20*
Sunwatch, *68*
swidden cultivation, 78, 84

TEK. *See* traditional ecological knowledge
teosinte, 41
tepary beans, 61
Three Sisters crop complex, 78, 80, 84
Timbisha Shoshone, 102
tobacco, 61, 81, 102
   native and European types of, 117
Toltec site, *68*, 73
trade, 86
   with Europeans, 106, 108
   Hopewell era, 35
   indirect, 106, 113, 116
   maize and, 37
traditional ecological knowledge (TEK), 123
trait associations, 13
tree nuts, 35
trinchera fields, 60
Tucson Basin, *39*
turkeys, vii, 58–59

ubiquity measure, 52–53
University of Michigan, 23
upland farming hypothesis, Woodland period EAC and, 32–34

Village Ecodynamics Project, 57
Vinette I, *68*, 69

wapato, 96
watermelon, 113, 114–115, 116, 117, 134

weeds (weedy crops). *See also specific plants*
   annuals, 12, 15
   EAC crops as, 15
   ecological definition of, 3
   environmental tolerance of, 4
   Eurasian, 111
   floodplain weed theory, 25–26, 127
   husbandry of, 48–49
   pre-maize cultivation of, 12
West Coast, unrecognized food production in, 9
Western Forests ecoregion, 6, 8–9
wheat, *112*, *115*, 116, 120, 144
whitetail deer, 7, 34–35, 86

wild plants, 3
Woodland period, 13, 24
   anthropogenic landscape effects and, 31–34
   farming communities, land tenure and, 34–35
   farming technology and, 32–34
   food production during, 28–36
   human diets in EAC in, 29–31, 37n2
   regional variability in EAC role during, 35–36
   social complexity and, 34–35

yellow glacier lily, 96